Higher Electronics

Higher Electronics

Mike James

Taylor & Francis Group

LONDON AND NEW YORK

First published by Newnes

First published 1999

This edition published 2011 by Routledge
2 Park Square, Milton Park, Abingdon, Oxon OX14 4RN
711 Third Avenue, New York, NY 10017, USA

Routledge is an imprint of the Taylor & Francis Group, an informa business

All rights reserved. No part of this publication may be reproduced in
any material form (including photocopying or storing in any medium by
electronic means and whether or not transiently or incidentally to some
other use of this publication) without the written permission of the
copyright holder except in accordance with the provisions of the Copyright,
Designs and Patents Act 1988 or under the terms of a licence issued by the
Copyright Licensing Agency Ltd, 90 Tottenham Court Road, London,
England W1P 9HE. Applications for the copyright holder's written
permission to reproduce any part of this publication should be addressed
to the publishers

British Library Cataloguing in Publication Data
A catalogue record for this book is available from the British Library

ISBN: 978-0-7506-4169-2

Library of Congress Cataloguing in Publication Data
A catalogue record for this book is available from the Library of Congress

Contents

Introduction   xi

1 Power supplies   1
  1.1 Rectification and smoothing   1
      Power rating   2
      RMS quantities   2
      Regulation   4
      Rectification   5
      Rectifier bridge   7
      Calculation of ripple suppression capacitor   8
  1.2 Voltage regulator diodes (Zener diodes)   10
      Design of simple regulators   11
  1.3 Linear discrete component voltage regulators   14
      The Darlington pair series pass element   16
      Overcurrent protection   17
  1.4 Designs using feedback techniques   18
      Variable output feedback regulator   20
      Variable output voltage with feedback and current limiting   20
      Using an op amp comparator   21
  1.5 Linear integrated circuit voltage regulators   22
      Fixed voltage three-terminal regulators   22
      Variable voltage three-terminal regulators   25
  1.6 Heatsinks   26
      Electrolytic capacitors   27
  1.7 Current limiting vs. foldback   28
  1.8 Voltage multipliers   30
      D.c. restoration   30
      Peak clamp   30
      Supply doublers   30
      Voltage level converters   31
      Hickman multiplier   31
  1.9 Constant current sources   32
      FET-based constant current sources   32
      BJT-based constant current circuits   33
      Op amp solutions   34
      IC voltage regulator   34
  1.10 Switched mode power supplies   34
      Buck converter   36
      The boost regulator   38
      The buck–boost regulator   38

	The flyback regulator	39
	More examples	39
	The charge pump	40

2 Feedback (1) — 41
- 2.1 Introduction — 41
- 2.2 Positive feedback — 43
- 2.3 Negative feedback — 44
- 2.4 Benefits of negative feedback — 46
  - Increase of bandwidth — 46
  - Reduction of amplifier noise — 49
  - Reduction of distortion — 50
- 2.5 Electronic mixer circuits — 52
  - The long tailed pair — 52
  - The base–emitter comparator — 55
  - The operational amplifier mixer — 56

3 Power amplifiers — 57
- 3.1 Types of amplifier — 57
  - Amplifier gain and frequency response — 57
  - Logarithmic axes — 59
  - The tuned amplifier — 59
  - Band pass, low pass and high pass amplifiers — 61
  - An audio problem, or the tweeter-buster — 63
- 3.2 Classes — 63
- 3.3 Class A amplifiers — 64
  - D.c. blocking — 65
  - Capacitive coupling — 66
  - Transformer coupling — 68
  - Class A push–pull design — 69
  - Darlington pairs — 70
  - Case study – the Linsley-Hood Class A amplifier — 73
- 3.4 Class B amplifiers — 75
  - Crossover distortion — 75
  - Class B output bias — 76
  - Class B case study — 77
  - Efficiency of Class B — 78
- 3.5 Class C amplifiers — 80
- 3.6 Integrated circuit power amplifiers — 80
  - Bridges — 81
  - Wiring — 83
  - Heatsinking — 83
  - Distortion — 84
  - Class B harmonic distortion — 85
  - Intermodulation distortion — 86

4 Feedback (2) — 87
- 4.1 Amplifier classification — 87
- 4.2 Series current — 88
  - An undecoupled emitter resistor ... series current feedback — 89
  - D.c. design — 91
  - JFET design — 93

	4.3	Series voltage	93
		The long tailed pair	95
	4.4	Shunt current	96
	4.5	Shunt voltage	98
		Designing a simple shunt voltage feedback amplifier	99
	4.6	Feedback problems	101

5 Operational amplifiers (1) — 102

- 5.1 Introduction — 102
  - Packages — 103
- 5.2 Designing with the ideal op amp — 104
  - Circuit 1 The buffer amplifier — 105
  - Circuit 2 The non-inverting amplifier — 105
  - Circuit 3 The inverting amplifier — 106
  - Circuit 4 The mixer or adder — 108
  - Circuit 5 The differential amplifier or subtractor — 109
- 5.3 Other operational parameters — 111
  - Bandwidth — 111
  - Slew rate limiting — 111
  - Multistage amplifiers — 115
- 5.4 Variable gain amplifiers — 116
  - Potentiometers — 117
- 5.5 Input offset errors — 117
  - Input offset current — 117
  - Input offset voltage — 117
- 5.6 Common mode rejection ratio — 118

6 Noise — 120

- 6.1 Interference — 120
- 6.2 Reduction of power supply interference — 120
  - Decoupling capacitors — 121
- 6.3 Ground loops — 122
  - Voltage sense circuits — 123
- 6.4 Crosstalk — 124
  - Capacitive (electrostatic) coupling — 124
  - Magnetic (inductive) coupling — 124
  - Coaxial cable screens — 125
- 6.5 Internal noise — 126
  - Johnson noise — 126
  - Shot noise — 129
  - Flicker noise or $1/f$ noise — 129
  - Burst (popcorn) noise — 130
- 6.6 Noise models — 130
- 6.7 Signal to noise ratio — 131
  - Noise figure (noise factor) — 131
  - Noise temperature — 132

7 Operational amplifiers (2) — 133

- 7.1 The instrumentation amplifier — 133
  - Bridge measurement methods — 136
- 7.2 Chopper stabilized amplifiers — 140
- 7.3 Comparators — 142

			Window detector	146
	7.4		Integrators	147
			D.c. performance	148
			Integrator applications	149
	7.5		Active filters	152
			Second order 'Sallen and Key' filters	155
			Sallen and Key filters with gain	159
			Notch filters	160
	7.6		Transconductance amplifiers	161
8	**Oscillators**			**163**
	8.1		The tuned collector oscillator	163
	8.2		Colpitts and Hartley oscillators	165
			Colpitts oscillators	167
			Hartley oscillators	167
	8.3		*RC* sinewave oscillators	169
			Phase shift ladder	169
	8.4		Wien Bridge oscillators	172
	8.5		Twin T oscillators	175
	8.6		Relaxation oscillators	177
			Resistor–capacitor timing	177
			The operational amplifier astable multivibrator	180
			The discrete transistor astable multivibrator	182
			Monostable multivibrators	185
	8.7		The 555 timer	187
			The 555 monostable multivibrator	188
			The 555 astable multivibrator	190
9	**Radio frequency and other techniques**			**192**
	9.1		Transistor circuits	192
			Miller effect	193
			The common base amplifier	194
			R.f. transformers	195
	9.2		Operational amplifiers	196
	9.3		Crystal oscillators	197
			Oscillators	199
			Ceramic resonators	200
	9.4		Phase locked loops	201
			Some common uses	202
			The component parts of a phase locked loop (PLL)	203
			The parameters of a phase locked loop	203
			The HEF 4046 PLL	207
			The programmable frequency synthesizer	208
10	**Logic circuits**			**211**
	10.1		Logic gates and manipulation	211
			Logic symbols	212
	10.2		TTL	212
			The NOT gate	212
			The 2-input NAND gate (Not-AND)	214
			The 2-input NOR gate (Not-OR)	214
			Output circuits	214
			IEC qualifying symbols	215
			Tri state logic	216
			Other IEC qualifying symbols	216

Families	216
Packages	217
Summary of good practice when working with TTL	218
10.3 CMOS	218
Electrostatic sensitivity	219
10.4 Combinational logic circuits	221
Inputs	221
10.5 Minimization	223
10.6 Programmable logic devices	225
Example solution	234

11 Sequential logic 238

11.1 Simple latches	238
The D-type latch	240
Truncated counts	241
11.2 Digital clock generators	244
Clock generator circuits	246
11.3 Synchronous design	246
J–K flip-flops	246
Master–slave vs. edge triggered operation	247
State diagrams	247
Synchronous counters	249
11.4 Shift registers	252
Universal shift registers	253
11.5 Programmable logic	253
11.6 Memories	255
Static RAM	255
Dynamic RAM	256
Read Only Memories	256

12 Microprocessors 259

12.1 Microprocessor fundamentals	259
12.2 The 8051 microprocessor	260
The watchdog timer	263
Clock and resonator circuits	264
Design example	267
12.3 Accessing external memory	267
Logical separation of program memory (read only) and data memory (read/write)	268
Non-volatile memory	269
Alternate use of Port 3	271
12.4 Internal architecture	271
Registers	271
Microprocessor instructions	272
Timer/counters	275
Interrupts	279
Stack	282
Serial communication	283
Baud rate generation	285
12.5 Analogue-to-digital and digital-to-analogue conversion	288
Digital-to-analogue converters (DACs)	288
Analogue-to-digital converters	292
Conversion methods	293

Appendix A	Number systems	297
Appendix B	MCS51 Instruction set	303
Index		307

Introduction

Electronics is a dynamic subject. The developments in components and techniques are exciting to follow and implement. This book has been written to get you up and started with real components and practical circuits as rapidly as possible, but without sacrificing the rigour which underpins the analysis of their performance. I would like to hope that an emphasis on the design exercises would give you the confidence to use electronics to solve problems and to apply it in whatever field of science you may be involved in.

Many students have problems with electronics as a subject and have difficulty 'getting into it'. The order in which the chapters are presented is designed to stimulate and relate real-world problems as soon as possible. Some areas of study are split; for example operational amplifiers. These represent an enormous range of components which are put to very diverse uses, and it seemed inappropriate to work through all of the applications all at once. It was split to present the fundamentals to start with (Chapter 5), and then, after another related topic, to press on into the more advanced uses of these ubiquitous devices (Chapter 7). In some cases, it was impossible to cover all of the subjects in the most appropriate location. So, for example, if you come into electronics from scratch, you will not be very familiar with small signal transistor amplifiers. These are used (and explained) in Chapter 3 (Power amplifiers), but are analysed more rigorously in Chapter 4 (Feedback 2).

Diagram conventions

Throughout the book, I have tried to adopt a common convention for the drawing of circuit diagrams. International standards call for the positive voltage supplies to be at the top of the diagram and the negative voltage supplies to be at the bottom. There are a few exceptions to this, and the allowable deviations occur in the section on power supplies. These follow the other convention which demands that inputs are on the left of the diagram and outputs on the right. There are a few exceptions even to this, but the component placement in the circuit should make it obvious what is intended.

Conductors should be placed orthogonally. That is, side-to-side and up/down. There are exceptions, but these are for fairly specific circuit elements such as Wheatstone measurement bridges and star connected earths.

Those of you who have experience of circuit diagrams from previous eras will recall how junctions and conductor cross-overs

Figure I.1

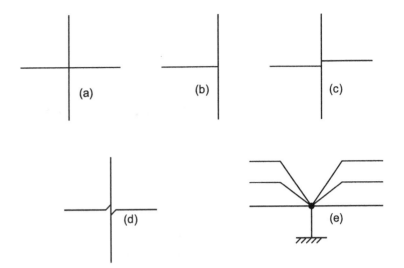

were handled. In this book, I will use the standards shown in Figure I.1:

(a) two wires crossing over but **not** electrically connected;
(b) a junction between three wires;
(c) a junction between four wires. Note the stagger in the alignment – this is essential if the circuit is not to be confused with a cross-over;
(d) a not so commonly seen variant of diagram (c);
(e) an exception: a star earth point.

What will **not** be seen is shown in Figure I.2:

(a) a junction;
(b) a cross-over.

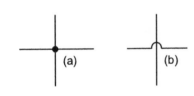

Figure I.2

In the book, I have tried to cover all the main topics which occur in modern electronics. It introduces many practical circuits which will serve in many applications. One section which is **not** covered is the use and application of software analysis tools. There are many of these and they are now an indispensable aid to the circuit designer, saving many hours of assembly and testing. They will analyse the performance of analogue, digital and mixed circuits, producing graphical or tabular output as required. Component manufacturers all make their component models available for inclusion in your designs (undoubtedly in the hope of increased sales). Software analysis was left out as a topic because there are a lot of these tools about and they are all used slightly differently.

My hope is that for whatever course you are following, then you will have access to them to support and test the circuits suggested. If not, try searching the Internet for Electronic Analysis and you should rapidly come across tools such as PSPICE, MicroSIM, $\langle\langle\rangle\rangle$, all of which are available for download in a cut-down evaluation version. These versions all work adequately, but are restricted in some way – usually in the number of components in the circuit.

The book has been written at a level which will be useful for Higher National Certificate and Diploma courses and at University standard. After sixteen years of teaching Electronics, I felt it was time to put some of that experience back in the form of an applied textbook which provides some of the analysis techniques taught, presented in an order which I felt worked and worked well in the classroom.

Mike James

1 Power supplies

Summary

This chapter deals with the problem of providing stable, smooth d.c. supplies that provide a constant voltage or current output. The technologies described cover the range of those in common use – linear, switch-mode, step-down, step-up, with limiting or foldback.

1.1 Rectification and smoothing

This section deals with the conversion of a.c. mains electricity into a low voltage d.c. supply. Later sections will deal with the design of precision d.c. regulators which provide an accurate, overcurrent-protected, voltage source.

The first step is to reduce the mains supply to a lower a.c. equivalent. A transformer is used for this purpose. The a.c. high voltage is connected to the primary windings while the lower a.c. output is available from the secondary windings. Figure 1.1.1 shows **three** possible configurations which are commonly available off the shelf. Most commercial equipment has to work at European and American a.c. transmission voltages, hence the variety of primary windings. Europe requires 230 V @ 50 Hz, while the USA requires 110 V @ 60 Hz.

The transformer core is made from steel laminations. It supports the primary and secondary windings and guides the magnetic flux

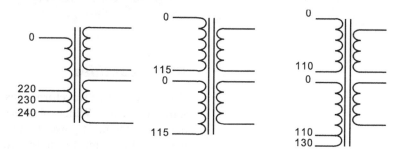

Figure 1.1.1 *Transformer configurations*

2 Higher Electronics

Table 1.1.1

6 VA 25% regulation	12 VA 12% regulation	50 VA 10% regulation
0–5 V 0–5 V @ 0.6 A	0–5 V 0–5 V @ 1.2 A	
0–6 V 0–6 V @ 0.5 A	0–6 V 0–6 V @ 1.0 A	0–6 V 0–6 V @ 4.17 A
0–9 V 0–9 V @ 0.33 A	0–9 V 0–9 V @ 0.67 A	0–9 V 0–9 V @ 2.78 A
0–12 V 0–12 V @ 0.25 A	0–12 V 0–12 V @ 0.5 A	0–12 V 0–12 V @ 2.08 A
0–15 V 0–15 V @ 0.2 A	0–15 V 0–15 V @ 0.4 A	0–15 V 0–15 V @ 1.67 A
0–18 V 0–18 V @ 0.17 A	0–18 V 0–18 V @ 0.33 A	0–18 V 0–18 V @ 1.39 A
0–20 V 0–20 V @ 0.15 A	0–20 V 0–20 V @ 0.3 A	0–20 V 0–20 V @ 1.25 A
0–24 V 0–24 V @ 0.12 A	0–24 V 0–24 V @ 0.25 A	0–24 V 0–24 V @ 1.04 A

Data: Farnell Electronic Components Ltd

which links them. At the frequencies of 50 Hz and 60 Hz, laminated cores provide the best cost/efficiency compromise. The two frequencies are close enough so that no modifications are needed to enable the transformer to operate with either.

There are usually **two** secondary windings. These can either be left separate, connected in series for a higher output voltage or connected in parallel to obtain a higher current.

Power rating

The transformer must also be able to deliver sufficient current for the job in hand. Transformers are rated in **volt-amperes** (VA), which are a measure of the amount of real power they can handle. Table 1.1.1 shows a typical list of mains transformers and their ratings.

The lower the power handling, the worse is the percentage regulation. This is a measure of output voltage drop which will be explained and allowed for in the next section. A 'dual 0–12 V 6 VA' transformer has two 0–12 V secondaries each capable of delivering $3\,\text{VA}/12\,\text{V} = 0.25\,\text{A}$. These could deliver a total of:

$$24\,\text{V} @ 0.25\,\text{A} \quad \text{or} \quad 12\,\text{V} @ 0.5\,\text{A} \quad \text{or} \quad 2 \times 12\,\text{V} @ 0.25\,\text{A}$$

Problems 1.1.1

What current is each secondary of the following transformers capable of delivering:

(1) dual 0–6 V 12 VA;
(2) dual 0–9 V 25 VA;
(3) dual 0–12 V 50 VA?

RMS quantities

Each of the three waveforms has the same peak amplitude and the same frequency. Which of these three do you think is capable of delivering the most power to a load? For instance, which waveform would light a lamp to the brightest extent?

Figure 1.1.2 *Voltage waveforms*

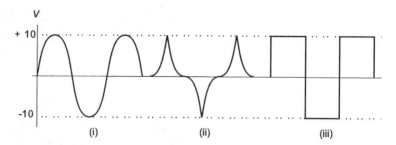

(i) (ii) (iii)

The amplitude is characterized by assigning an amplitude quantity called V_{RMS} to it. RMS stands for 'root mean square' and it describes the mathematical process by which the voltage is calculated. For the waveforms above:

$$V_{RMS}(i) = 7.07 \text{ V}; \quad V_{RMS}(ii) = 2.00 \text{ V}; \quad V_{RMS}(iii) = 10.00 \text{ V}$$

The power delivered by each of these three is given by:

$$\text{Power} = V_{RMS} \times I_{RMS} \quad \text{or} \quad \text{Power} = \frac{V_{RMS}^2}{R} \quad \text{or} \quad \text{Power} = I_{RMS}^2 \times R$$

Since a.c. is generated as a sine wave, then this is the figure which is most relevant. For a sine wave, it is fairly easy to prove that:

$$V_{RMS} = \frac{V_{PEAK}}{\sqrt{2}}$$

Mathematics in action

All of these waveforms are symmetrical, so their average value over one cycle will be **zero**. However, when you square a negative number, the answer is positive: for example, $-2 \times -2 = +4$. This is the basis of calculating the root mean square. For any periodic waveform:

> **Square** the value at all instants
> Find the **mean** value
> Take the **root** of this mean of the squares.

Hence **root-mean-square**.

When a sine wave has a value of $v = V_{PK} \sin wt$ the square is $v^2 = V_{PK}^2 \sin^2 wt$. The mean value over one period is found by integrating the expression with respect to time from the range 0 to 2π. This gives the area under the square of the waveform. The average is then:

$$\text{average} = \frac{\text{area}}{2\pi}$$

Working through this:

$$\text{area} = \int_0^{2\pi} V_{PK}^2 \sin^2 wt \, dt \quad \text{where } \sin^2 wt = \tfrac{1}{2} - \tfrac{1}{2} \cos 2wt$$

$$= \int_0^{2\pi} (\tfrac{1}{2} - \tfrac{1}{2} \cos 2wt) \, dt$$

$$= \frac{V_{PK}^2}{2}[t - \tfrac{1}{2}\sin 2wt]_0^{2\pi}$$

$$= \frac{V_{PK}^2 2\pi}{2}$$

$$\text{mean value} = \frac{\text{area}}{2\pi}$$

$$\text{mean} = \frac{V_{PK}^2}{2}$$

The **root** of the **mean** of the **squares** is thus:

$$V_{RMS} = \frac{V_{PK}}{\sqrt{2}}$$

Problems 1.1.2

(1) Calculate the peak voltage and the peak–peak voltage of the UK 230 V_{RMS} mains supply.
(2) What is the peak voltage to be expected from the secondary of a:
  (a) 12 V mains transformer
  (b) 15 V mains transformer

(transformer voltages are always quoted as RMS values).

Regulation

All real transformers exhibit a property called regulation. In a 12 V, 0.5 A secondary, the secondary will only output 12 V when it is actually delivering 0.5 A. At values of current less than this, the RMS voltage output will be higher, as shown in Figure 1.1.3.

When zero current is drawn from the transformer, the secondary voltage rises to 13.44 V (in this case). The figure of 13.44 V comes from the fact that different VA ratings of transformers have the typical regulation figures shown in Table 1.1.2.

Table 1.1.2

VA rating	6	12	20	50	75	100	150	200
% Regulation	25	12	10	10	10	10	10	7

Data: Farnell Electronic Components Ltd

So, from Tables 1.1.1 and 1.1.2, for a 12 VA transformer, the no-load secondary RMS voltage would be 12% higher or 13.44 V as shown. This effect is fairly linear, so a load of 0.25 A would result in a secondary voltage of 12.72 V.

Figure 1.1.3 *Output voltage/current characteristic of a transformer*

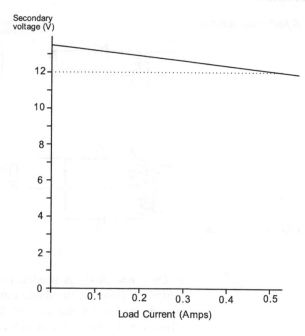

There is a general rule that the smaller the transformer, the less magnetically efficient it is and so the worse the regulation. The leakage flux does not provide perfect linkage between primary and secondary windings. Another factor is the I^2R losses which increase as the wire sizes get smaller. For example, the miniature encapsulated transformer range has the figures shown in Table 1.1.3.

Table 1.1.3

VA rating	2	4	6	10	14	18	25	30
% Regulation	34	34	31	28	26	23	23	19

Data: Farnell Electronic Components Ltd

Problems 1.1.3

(1) (a) What is the no-load secondary voltage of a practical 15 V, 4 VA encapsulated transformer as in Table 1.1.3?
 (b) What would the secondary voltage be when delivering 100 mA?
(2) What is the secondary terminal voltage of a 9 V, 6 VA transformer delivering 250 mA? (Use data from Table 1.1.1.)

Rectification

The next step is to convert the alternating voltage into a unidirectional voltage. Rectifier diodes are used for this.

6 Higher Electronics

Figure 1.1.4 *A half wave rectifier*

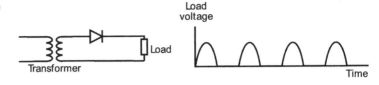

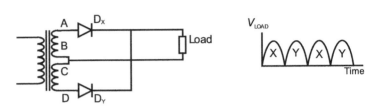

Figure 1.1.5 *A full wave rectifier*

Only **one** diode is needed for a half wave rectifier. When the output is positive, the diode conducts. When the voltage reverses, the diode blocks the current to produce the waveform shown. However, because silicon diodes are so cheap, it is rarely worth using this inefficient method. Full wave rectification can be achieved by one of two methods (Figures 1.1.5 and 1.1.6).

Diodes X and Y alternately conduct as secondary terminals A and D change polarity every half cycle. The load voltage will be $(V_{AB} \times \sqrt{2} - 0.6\,\text{V})$ or $(V_{CD} \times \sqrt{2} - 0.6\,\text{V})$ depending on the mains polarity at the time.

$(V_{AB} \times \sqrt{2})$ and $(V_{CD} \times \sqrt{2})$ is the peak output; 0.6 V is the forward voltage drop across each diode as it conducts.

Example 1.1.1

Calculate the peak load voltage of an ideal (regulation-less) 12 V transformer.

$$\text{Peak secondary voltage} = 12 \times \sqrt{2}$$
$$= 17\,\text{V}$$
$$\text{Peak load voltage} = 17\,\text{V} - 0.6\,\text{V}$$
$$= 16.4\,\text{V}$$

Problems 1.1.4

(1) Calculate the peak load voltage of a rectified ideal 9 V transformer.
(2) Calculate the peak load voltage of a rectified 2 VA 0–6/0–6 transformer when delivering 200 mA.
(3) What is the maximum d.c. voltage expected from a rectified 15 V 50 VA transformer?

Figure 1.1.6 *A full wave rectifier bridge*

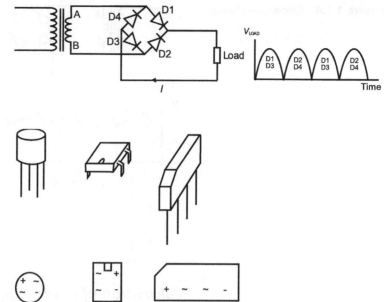

Figure 1.1.7 *Typical bridge rectifier packages*

Rectifier bridge

One of the most common configurations used for full wave rectifying an a.c. waveform is the rectifier bridge. Note that the direction of the diodes D1 and D2 point up to the +d.c. while D3 and D4 point away from the −d.c. (Figure 1.1.6).

The diodes conduct in pairs:

When A is more positive than B, diodes D1 and D3 will conduct.

When B is more positive than A, diodes D4 and D2 will conduct.

Since there are always **two** diodes in the circuit path, the peak load voltage is 1.2 V less than the peak transformed voltage!

In normal use, rectifier bridges are integrated into a single package. These generally work out cheaper and are more compact than four discrete diodes (Figure 1.1.7).

Problems 1.1.5

(1) Calculate the peak load voltage of a bridge rectified (ideal) 15 V transformer.
(2) What is the peak load voltage of a bridge rectified (ideal) 3 V transformer?
(3) The two secondaries of a 0–9 0–9 6 VA transformer are connected in series:
   (a) What would be the peak output voltage from a bridge rectifier with no load?
   (b) What would be the peak output voltage from a bridge rectifier when delivering 200 mA?

8 Higher Electronics

Figure 1.1.8 *Ripple waveforms*

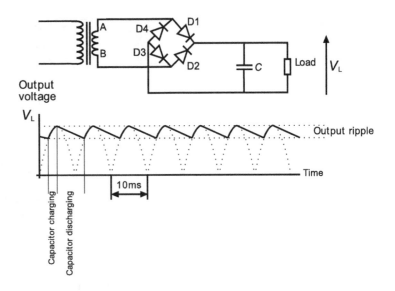

Calculation of ripple suppression capacitor

The full wave rectified waveform is unidirectional (positive only or negative only, dependent on how the rectifier is connected). In countries where the mains frequency is 50 Hz, the dips fall to zero at 10 ms intervals because this mains frequency has a periodic time of 20 ms. Each half wave will thus take 10 ms. To 'hold up' the waveform during the supply dips, a capacitor is used to maintain an approximately constant output. Figure 1.1.8 shows the effect on the output waveform when a capacitor is used to store energy at the voltage peaks ready to release it when the voltage falls.

The voltage represented by this waveform is usually fed into a regulator circuit. This will maintain a precise and controlled supply for whatever load is used. However, the main ripple capacitor C is calculated only from a knowledge of the load current. Recall that for a capacitor:

$$Q = C \times V$$

where

$$I \times t = C \times V$$

I = load current (Amps)

t = ripple period (Seconds)   10 ms – full wave rectification
                                20 ms – half wave rectification

C = capacitance (Farads)

V = ripple voltage (Volts)

Example 1.1.2

Calculate a suitable capacitor for smoothing a full wave rectified signal to a 1 A load if the ripple voltage is not to exceed 0.5 V.

$$I \times t = C \times V$$

$$1\,A \times 10\,ms = C \times 0.5\,V$$

from which:

$$C = 20\,000\,\mu F$$

Be aware that the typical tolerance for electrolytic capacitors is of the order of 20%. They are also rated by voltage and so when choosing a capacitor, ensure that it is adequate for the job in hand. Typical values are:

10	22	33	47	µF
100	220	330	470	µF
1 000	2 200	3 300	4 700	µF
10 000	22 000	33 000	47 000	µF

at voltage ratings of:

16	25	40	63 Volts

This represents a typical range of values – other values are available at other voltage ratings. In general, the higher the capacitance and the higher the voltage rating, then the higher the cost and the larger the physical package size.

Problems 1.1.6

(1) Calculate a suitable capacitor to smooth a 0.4 A full wave rectified supply to a load so that the ripple does not exceed 0.75 V.

(2) A certain load draws 1.8 A from a full wave rectified supply. What value smoothing capacitor should be used if the voltage ripple is not to exceed 2.5 V?

(3) A full wave rectified supply is smoothed by a 1000 µF capacitor. What is the maximum current which can be drawn if the ripple voltage is not to exceed 1 V?

(4) The output of a 9 V mains transformer is full wave rectified, smoothed and delivers 1.5 A to its load. Calculate:
   (a) peak voltage output;
   (b) ripple voltage if the smoothing capacitor is 10 000 µF;
   (c) the maximum and minimum voltages of the waveform delivered to the load.

(5) The dual 15 V secondaries of a mains transformer are connected with two diodes to provide a full wave rectified output. What value of smoothing capacitor is required so that the ripple voltage does not exceed 3.3 V when delivering a current of 2.1 A? Sketch a suitable circuit diagram.

(6) A 6 V mains transformer has its output rectified and smoothed and is expected to deliver up to 250 mA to a load. What value of smoothing capacitor would be required if the load voltage is never to fall below 7.0 V?

Figure 1.2.1 *Reverse breakdown*

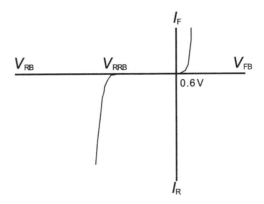

1.2 Voltage regulator diodes (Zener diodes)

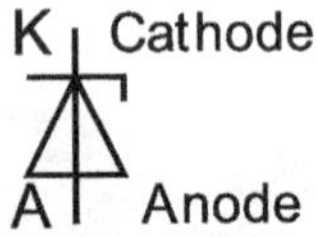

Figure 1.2.2 *A Zener diode*

If a reverse biased diode is submitted to a voltage beyond its peak inverse voltage (PIV) rating, it will suddenly (and usually catastrophically) start conducting. This is the point V_{RRB} – the reverse breakdown voltage of Figure 1.2.1.

This effect can be put to a beneficial use. By careful control of doping levels and diode geometry, the reverse breakdown can be designed to (almost) any level. The device is then called the Zener or avalanche diode. These useful devices are central to most linear voltage regulator circuits.

When forward biased, they act as conventional silicon diodes, but are mostly valued for the controlled and repeatable breakdown characteristic.

There are in fact two mechanisms which can cause the diode to suddenly start conducting when reverse biased – the Zener effect and the avalanche effect. The Zener effect occurs below approximately 5.6 V and the avalanche effect occurs above approximately 5.6 V.

The two mechanisms exhibit different temperature coefficients as shown. Thus it is possible to find a point on a 5.6 V breakdown diode where the voltage is stable, regardless of temperature. Despite the two different effects being evident, both devices are colloquially referred to as 'Zener diodes'.

Zener diodes are rated by breakdown voltage and power handling capability. Thus, from the table, a BZX79C10 is a 10 V 400 mW Zener diode.

$$P = V \times I \quad \text{so} \quad 400\,\text{mW} = 10 \times I$$

which means that this device can handle a maximum of 40 mA. In

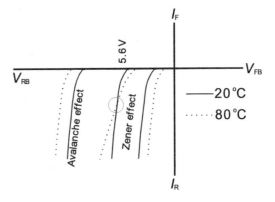

Figure 1.2.3 *Zener and avalanche diode characteristics*

Power supplies

Table 1.2.1 *Typical Zener diodes*

V_{RRB}	400 mW	500 mW	1.3 W
2.4	BZX79C2V4	BZX55C2V4	
2.7	BZX79C2V7	BZX55C32V7	
3.0	BZX79C3V0	BZX55C3V0	
3.3	BZX79C3V3	BZX55C3V3	1N5333B
3.6	BZX79C3V6	BZX55C3V6	
3.9	BZX79C3V9	BZX55C3V9	1N5335B
4.3	BZX79C4V3	BZX55C4V3	1N5336B
4.7	BZX79C4V7	BZX55C4V7	1N5337B
5.1	BZX79C5V1	BZX55C5V1	1N5338B
5.6	BZX79C5V6	BZX55C5V6	1N5341B
6.2	BZX79C6V2	BZX55C6V2	1N5342B
6.8	BZX79C6V8	BZX55C6V8	1N5343B
7.5	BZX79C7V5	BZX55C7V5	1N5344B
8.2	BZX79C8V2	BZX55C8V2	1N5345B
9.1	BZX79C9V1	BZX55C9V1	1N5346B
10	BZX79C10	BZX55C10	1N5347B
11	BZX79C11	BZX55C11	1N5348B
12	BZX79C12	BZX55C12	1N5349B
13	BZX79C13	BZX55C13	1N5350B
15	BZX79C15	BZX55C15	1N5352B
16	BZX79C16	BZX55C16	1N5353B
18	BZX79C18	BZX55C18	1N5355B
20	BZX79C20	BZX55C20	1N5357B
22	BZX79C22	BZX55C22	1N5358B
24	BZX79C24	BZX55C24	1N5359B
27	BZX79C27	BZX55C27	1N5361B
30	BZX79C30	BZX55C30	1N5363B

use, the device is derated from this maximum so that it does not become hot and electrically stressed, which would reduce its working life.

Problems 1.2.1

Find the maximum current handling of:

(a) BZX79C5V6
(b) BZX55C7V5
(c) 1N5355B
(d) BZX79C2V7
(e) BZX55C15
(f) 1N5343B.

Design of simple regulators

Most supplies in this category take the form of the circuit of Figure 1.2.4. The input is an unregulated voltage which is 'clamped' by the Zener diode. The series resistor R_S drops the voltage from V_{IN} to V_Z. The art of designing one of these regulators is to calculate the size and rating of R_S.

Figure 1.2.4 *A simple Zener regulator*

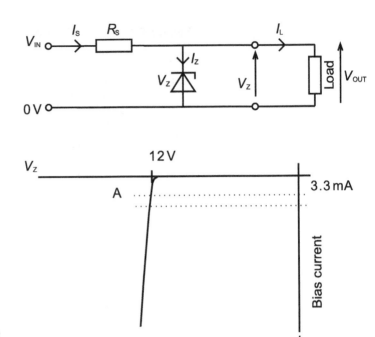

Figure 1.2.5 *A 12 V Zener diode characteristic*

Example 1.2.1

Design a simple 12 V Zener stabilized regulator which provides 100 mA to a load from a 17 V unregulated input.

The first steps are to decide on a Zener diode and then to set its current bias value.

Choose a BZX79C12, for instance. This 400 mW device can pass an absolute maximum of 33 mA (400 mW/12 V). This needs to be derated to give a margin of safety. Derating to 10% should be adequate, and implies that the Zener diode should operate at 3.3 mA.

Because of the steep slope of the characteristic at point A, a change of bias current alters the Zener voltage only slightly. It is a fact that most Zeners will operate quite satisfactorily at a bias current of 5 mA, so it is quite an interesting design procedure to bias **all** Zeners initially at 5 mA and then to check to see that the power rating is not compromised.

R_S has to supply 100 mA + 5 mA = 105 mA, and has to drop 17 V − 12 V = 5 V.

Thus, R_S has a value of 5 V/105 mA = 47.6 Ω and a rating of 5 V × 105 mA = 525 mW. The nearest preferred values would be a 47 Ω, 1 W resistor.

Example 1.2.2

Design a 6.2 V regulator to provide a load with 250 mA from a 9 V unregulated supply.

Figure 1.2.6 shows how the circuit would be laid out.

Choose a BZX79C6V2. Let the bias current be 5 mA.

Figure 1.2.6 *A Zener regulator circuit*

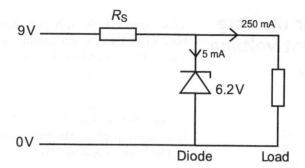

(Check: 5 mA × 6.2 V = 31 mW – well within the power handling capability.)

R_S has to deliver 250 mA + 5 mA = 255 mA and has to drop 9 V − 6.2 V = 2.8 V.

Thus, R_S has a value of 2.8 V/255 mA = 11 Ω and a rating of 2.8 V × 255 mA = 714 mW.

Disadvantages

There is a major disadvantage to this type of voltage regulator. If the load becomes disconnected, the 250 mA load current has to flow through the Zener diode. The Zener power dissipation becomes 1.581 W (which is 6.2V × 255 mA). This would lead to destruction of the device. Hence in practice this type of regulator should only be used where there is no possibility of the load being disconnected. The other disadvantage is the relatively high power dissipation of the series limit resistor.

Problems 1.2.2

(1) Design a voltage regulator which maintains a load at 7.5 V while delivering 150 mA from a 12 V unregulated source.

(2) A certain load needs to be held at 5.1 volts @ 85 mA. If a 10.5 V unregulated supply is available, design a suitable regulator.

(3) A 12 V 1 W lamp has to be operated from an 18 V unregulated supply:
   (a) Design a suitable zener stabilized supply.
   (b) Explain why this may not be the most suitable solution.

(4) A certain personal stereo requires approximately 3 V whilst drawing 112 mA. Design a system for powering it from a 9 V PP3 battery. What will be the total power drain on the PP3?

1.3 Linear discrete component voltage regulators

This section deals with very simple voltage regulators which are built from discrete semiconductor devices. They will not perform as well as the monolithic integrated circuit regulators to be considered in Section 1.5, but they do give a good insight into the basic properties and structure of linear regulator technology.

First, a revision of the basic properties of transistors. We shall use npn bipolar transistors exclusively. They are cheap, robust and have easily predictable performance.

The basic properties are:

- The transistor will not start conducting until the base–emitter voltage is approximately 0.6 V.
- When this happens, the current flowing from the collector terminal to the emitter terminal is controlled by the current entering the base terminal.
- The total emitter current is $I_E = I_C + I_B$.
- The transistor exhibits a current gain called h_{FE}, which is the ratio of I_C to I_B.
- Typically h_{FE} is:

    300 for signal transistors ($I_C \sim 100$ mA) and

    50 for power transistors ($I_C > 500$ mA).

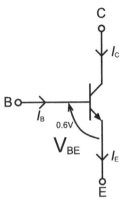

Figure 1.3.1 *npn transistors*

Problems 1.3.1

(1) If h_{FE} is 250 and I_B is 50 µA, what is I_C?
(2) What value of base current is required to cause a 115 mA collector current to flow in a transistor with h_{FE} of 175?
(3) If I_C is 50 mA and I_B is 0.4 mA what is h_{FE}?
(4) A transistor with base current of 75 mA and h_{FE} of 120 would produce how much collector current?

It must also be realized that h_{FE} is not a constant value. It varies with temperature, signal frequency, and value of I_C. With any batch of transistors, there is also a natural spread of h_{FE} within any sample. For example, the BC107 has its h_{FE} quoted as 115–450.

In the classic series pass transistor configuration (Figure 1.3.2), the voltage reference is set in the usual way by Zener diode Z1. It

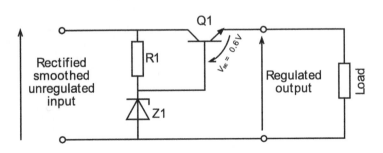

Figure 1.3.2 *The series pass transistor*

Figure 1.3.3

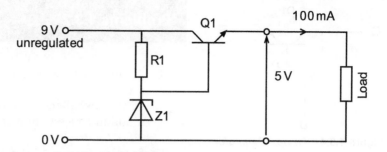

holds the base of Q1 at a fairly constant value. The voltage across the load will be:

$$V_{\text{LOAD}} = Z1 - 0.6 \text{ V}$$

The 0.6 V comes of course from the V_{BE} voltage drop across the transistor.

Example 1.3.1

Design a 5.0 V voltage regulator to deliver 100 mA to a load from a 9 V unregulated supply.

The first step would probably be to choose and set up the Zener bias circuit. Since the output needs to be 5.0 V, then choose a BZX79C5V6. This 400 mW Zener diode has a forward voltage drop of 5.6 V. The 5.6 V at the base of the transistor becomes 5.0 V at the emitter.

Let the Zener diode be biased at 10 mA. (Check: 10 mA × 5.6 V = 56 mW – OK.) This is well within the 400 mW limit. Choose a ZTX 300 transistor. This has $I_{\text{C MAX}}$ of 500 mA and h_{FE} of ~50.

If I_C is to be 100 mA, then I_B must be 100/50 = 2 mA. Thus the total current in $R1$ would be 12 mA. Should the load be removed, then Z1 would have to handle that 12 mA. 12 mA × 5.6 V = 67.2 mW which is still well within the maximum rating of the Zener diode.

$R1$ should be $(9 - 5.6)$ V/12 mA = 283 Ω @ $(9 - 5.6)$ V × 12 mA = 41 mA.

There could be literally hundreds of transistors from which to choose, but here is a selection:

Type	$I_{\text{C MAX}}$ A	h_{FE} min	P_{TOT} W
2N5321	2.0	30	1.0
2N2102	1.0	40	1.0
BC635	1.0	40	1.0
ZTX300	0.5	50	0.3
BC337	0.5	100	0.8
ZTX450	1.0	100	1.0
ZTX650	2.0	100	1.0

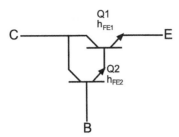

Figure 1.3.4 *An npn Darlington pair*

Problems 1.3.2

(1) Design a 10.0 V voltage regulator to deliver 250 mA to a load from a 16 V unregulated supply. (*Hint*: use a 10 V Zener diode, but put an ordinary silicon signal diode in series with it to increase the base voltage to 10.6 V.)
(2) Starting from a 12 V mains transformer, design a 9 V battery eliminator capable delivering up to 400 mA. (*Hint*: a 3.9 V Zener in series with a 5.6 V Zener gives approximately the correct voltage.)

The Darlington pair series pass element

The Darlington transistor configuration is used to improve the current gain of the transistor stage.

Transistor Q2 amplifies the current into its base. So $I_{C2} = h_{FE2} \times I_{B2}$. Its emitter is connected into the base of Q1. This current is then amplified once more so that $I_{C1} = h_{FE1} \times I_{B1}$. This provides an overall current gain which is theoretically $h_{FE1} \times h_{FE2}$, but is more typically quoted as ~1000.

The device could be made from two separate transistors, but is also available in a single transistor package with the conventional three pins base, collector and emitter.

Type	$I_{C\,MAX}$ A	h_{FE} min	P_{TOT} W
MPSA29	0.5	10 000	0.625
BCX38A	0.8	1 000	1.0
BCX38C	0.8	10 000	1.0
ZTX603	1.0	2 000	1.0
TIP110	1.0	1 000	50
BD677	1.5	750	40
TIP121	3.0	1 000	65
TIP131	4.0	1 000	70
TIP141	5.0	1 000	125
2N6059	6.0	750	150

Example 1.3.2

Design a power supply which is capable of delivering 15 V at 1.5 A to a load. Specify all components.

The simplest way of adjusting the output of this type of regulator is to be able to set the Zener diode to any voltage on demand (Figure 1.3.5).

In this circuit, the variable resistor *VR*1 is being used to tap into a percentage of the Zener voltage. There is a potential problem with this circuit. The base of Q1 must not take significant amounts of current compared with the current flow-

Figure 1.3.5 *A simple variable voltage supply*

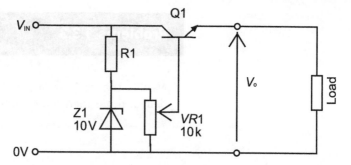

ing through VR1. If it does so, then the output will not be as expected.

For example, if Z1 = 10 V and VR1 = 10k, then simple application of Ohm's Law will tell you that the current through VR1 is 1 mA. If Q1 demands more current than this, then problems will occur.

In practice, this is not the best method of gradually controlling the output voltage. This will be illustrated in Section 1.4.

Overcurrent protection

Most overcurrent protection schemes rely on detecting the voltage drop across a sense resistor. The simplest scheme is shown in the circuit of Figure 1.3.6. This uses the fact that the transistor will turn on when the V_{BE} volt drop approaches 0.6 V, i.e. when $I_{OUT} = 0.6/R_S$, and current will flow from collector to emitter. This control (collector) current can be used to shut down another part of the circuit.

There are two ways of interpreting the control action:

- As Q2 turns on, it 'robs' the base current of Q1 by offering a lower impedance path for that current.
- When a transistor is fully 'on', its collector–emitter voltage drops to < 0.1 V. In doing so, it 'shorts-out' the base–emitter junction of Q1 and prevents it from operating properly.

Example 1.3.3

Design an overcurrent device which will prevent the power supply from delivering more than 0.5 A.

The transistor Q2 must start to turn on when a current of 0.5 A passes through the series limit resistor. R_S must have a value of 0.6 V/0.5 A = 1.2 Ω.

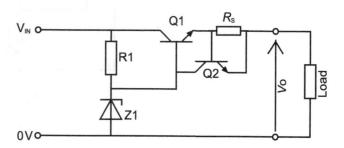

Figure 1.3.6 *Overcurrent protection*

Problems 1.3.3

Modify the designs of Problems 1.3.2 so that they cannot deliver more than the specified currents.

1.4 Designs using feedback techniques

Feedback is an important technique because it allows the actual output to be compared with the reference voltage. If there is any difference, then corrective action can be taken.

This is a completely different sort of circuit to that so far encountered. The output resistors $R3$ and $R4$ present a fixed proportion of the output voltage at the base of Q2. When all is working well, the base–emitter voltage across Q2 will be 0.6 V. If V_o rises or falls by the slightest amount, then the conductivity of Q2 is affected. Q2 takes the function of the comparator. It compares the output voltage (or that proportion determined by $R3:R4$) with the reference voltage V_z set by Z1. The output voltage is set by:

$$V_z + 0.6 = V_o \frac{R4}{R3 + R4}$$

To understand the corrective action, it is necessary to go back over the operation of a transistor:

- When the base–emitter voltage rises above 0.6 V, the collector–emitter voltage decreases.
- When the base–emitter voltage falls below 0.6 V, the collector–emitter voltage increases.

If, for some reason, the output voltage falls by a very small amount:

- the voltage on the base of Q2 will fall   which causes
- the collector–emitter voltage to rise   which causes
- the base voltage of Q1 to rise   which causes
- the output voltage (emitter voltage of Q1) to rise, correcting the original change.

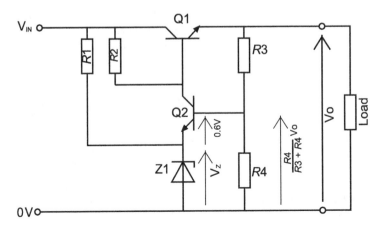

Figure 1.4.1

Figure 1.4.2 *The complete design*

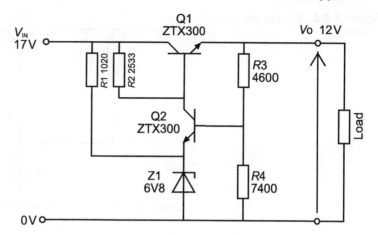

The other components:

- Resistor $R1$ provides the current drive into the base of Q1.
- Resistor $R2$ biases Z1.

Example 1.4.1

Design a voltage regulator capable of providing 100 mA at 12 V from a 17 V supply.

Choose a 6.8 V Zener diode.

Let its bias current be 10 mA. (6.8 V × 10 mA = 68 mW – OK.)

Design the circuit so that $R1$ provides 9 mA and $R1$ 1 mA. (Q2 will need some current through it to operate properly.)

Calculate $R1$ as $(17 - 6.8)/9\,\text{mA} = 1020\,\Omega$.

If a ZTX300 transistor is used, the gain is 50 (min) and so the base current would be 2 mA.

Calculate $R2$ as $(17\,\text{V} - 9.4\,\text{V})/(2\,\text{mA} + 1\,\text{mA}) = 2533\,\Omega$.

Q2 has an I_C of 1 mA. If the same ZTX300 is used the I_B of Q2 will be 20 μA.

The current flowing through $R3 : R4$ must be significantly bigger than 20 μA.

Let I_{R3R4} be 1 mA.

The base voltage of Q2 should be 6.8 V + 0.6 V = 7.4 V.

Calculate $R3$ as $(12\,\text{V} - 7.4\,\text{V})/1\,\text{mA} = 4600\,\Omega$.

Calculate $R4$ as $7.4\,\text{V}/1\,\text{mA} = 7400\,\Omega$.

$$V_z + 0.6 = \frac{R4}{R3 + R4} V_o$$

$$V_o = \frac{R3 + R4}{R4}(V_z + 0.6)$$

Note the alternate feedback circuit which allows the output to be variable.

Figure 1.4.3 *Varying the output voltage*

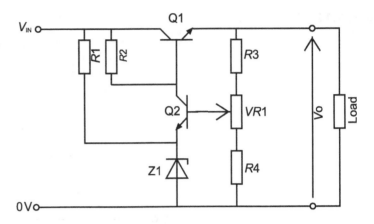

Problems 1.4.1

(1) Design a feedback stabilized circuit which provides a 4.5 V output at 1 A from an 11 V unregulated supply.
(2) Design a regulator which will output 6 V at 500 mA from an 18 V unregulated supply.
(3) Invert the circuit design to provide −8 V at 300 mA from a −13 V supply.

Variable output feedback regulator

In the circuit of Figure 1.4.3, the feedback point is derived from the wiper of the potentiometer. If:

$$R3 = 1k$$
$$VR1 = 1k$$
$$R4 = 1k$$

then when the wiper is at the bottom of $VR1$ the ratio would be:

$$\frac{1+1+1}{1}(V_z + 0.6) = \frac{3}{1}(V_z + 0.6)$$

When the wiper is at the top of $VR1$, the ratio would be:

$$\frac{1+1+1}{1+1}(V_z + 0.6) = \frac{3}{2}(V_z + 0.6)$$

From this, you can calculate the variability of the output.

Variable output voltage with feedback and current limiting

The circuit has now been modified to provide a current limit facility as described in Section 1.3. Note that the current limit is placed before the feedback resistors. This means that it is the actual output voltage which is being measured. Any voltage drop due to R_S is compensated for.

Figure 1.4.4 *Current limiting*

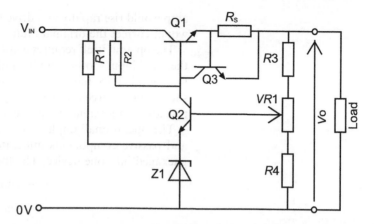

Problems 1.4.2

(1) Design a regulator to give an adjustable output of between 8 V and 10 V from an unregulated 15 V supply and which will provide up to, but not more than, 0.25 A output.
(2) A regulator is required to provide 5 V ± 0.5 V adjustment at up to 0.75 A. Design a suitable circuit.
(3) Design a voltage regulator to output between 13 V and 17 V at no more than 400 mA, derived from a 25 V unregulated supply.

Using an op amp comparator

Although this section was intended to look at discrete component regulators only, the use of the operational amplifier is included because it is a common improvement on the circuit performance (Figure 1.4.5).

This is a similar circuit to Figure 1.4.2 in terms of its functionality, but because of the higher gain of the op amp, the regulation is actually better. The op amp is connected so that the output changes to correct any deviation of V_X from the value of Z1. If, for some reason, V_{OUT} were to fall, then V_X would fall. The output of the op

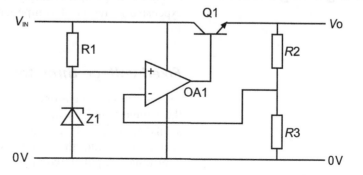

Figure 1.4.5 *Voltage regulator with op amp error amplifier*

amp would rise rapidly and drive the base of Q1 more positive. This would correct the original fault.

The op amp itself requires a voltage supply and this must be from the 'raw' input – otherwise it would never start up when power was first applied to the circuit. If used, the overcurrent protection should be between the series pass transistor and the feedback resistors in the same manner as Figure 1.4.4.

The operational amplifier is a circuit comprised of many active and passive components integrated onto one piece of silicon and packaged into one device. The five essential connections provide:

$$+ \text{signal in}$$
$$- \text{signal in}$$
$$\text{signal output}$$
$$+ V \text{ supply}$$
$$- V \text{ supply}$$

The op amp has a very high gain of typically 1 000 000, so that when two voltages are applied to the signal inputs:

$$V_o = 1\,000\,000(+V_{in} - -V_{in})$$

This improves the sensitivity of the power supply compared with that of Figure 1.4.1.

Problems 1.4.3

(1) Design an op-amp-based voltage regulator which provides a stable 13.5 V output at 750 mA from an unregulated 21 V input.
(2) Perform a complete mains-to-stable-voltage-regulator design. The device should provide 12 V out at 3 A and be overcurrent protected. It should be adjustable over the range of 11–13 V only.

1.5 Linear integrated circuit voltage regulators

In new designs, most linear power supply units tend to be based on IC regulators. Discrete component regulators are still designed, but only for very specialist applications – usually involving high current or high voltage. Quite simply, discrete component PSUs are more expensive in terms of component cost, assembly cost and circuit size.

Fixed voltage three-terminal regulators

Three-terminal regulators can provide a fixed output and represent the simplest and cheapest solution to most problems. They are available in all of the common voltage outputs (with an accuracy of around 4%) and the configuration can be adapted to give an adjustable output if required. This requires the use of additional

Figure 1.5.1 *Three-terminal regulator pinout*

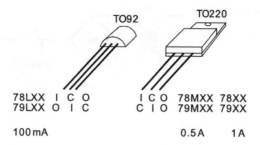

external components. Table 1.5.1 shows a selection of available devices.

All of these are overcurrent protected and have additional circuitry to prevent overheating. The 78xx and 79xx families are sourced by many manufacturers and are the de facto standard for the designer who wants the simplest, low component count regulator. One strong advantage of using them is that the supply for each part of a complex circuit can be derived and regulated locally rather than being dependent on a distributed system with its inherent voltage drops and electrical noise routing ability.

The application sheets show Figure 1.5.2 as a typical positive voltage supply. Of course, there are a few constraints:

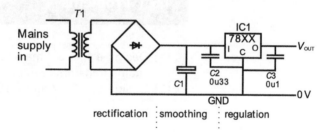

Figure 1.5.2 *Three-terminal regulator application circuit*

Table 1.5.1

Device	Input	o/p	Line R	Load R	o/p noise	Package
78L05	7–30 V	5 V @ 0.1 A	0.36%		40 μV	TO92
78L12	14.5–35 V	12 V @ 0.1 A	0.25%		80 μV	TO92
78L15	17.5–35 V	15 V @ 0.1 A	0.25%		90 μV	TO92
79L05	−7 to −30 V	−5 V @ 0.1 A	0.36%		40 μV	TO92
79L12	−14.5 to −35 V	−12 V @ 0.1 A	0.25%		80 μV	TO92
79L15	−17.5 to −35 V	−15 V @ 0.1 A	0.25%		90 μV	TO92
78M05	7–30 V	5 V @ 0.5 A	0.2%	0.4%	40 μV	TO220
78M12	14.5–35 V	12 V @ 0.5 A	0.84%	0.2%	75 μV	TO220
78M15	17.5–35 V	15 V @ 0.5 A	0.9%	0.17%	90 μV	TO220
79M05	−7 to −30 V	−5 V @ 0.5 A	0.7%	1.5%	125 μV	TO220
79M12	−14.5 to −35 V	−12 V @ 0.5 A	0.9%	0.55%	300 μV	TO220
79M15	−17.5 to −35 V	−15 V @ 0.5 A	0.9%	0.45%	375 μV	TO220
7805	7–30 V	5V @ 1 A	0.3%	0.2%	40 μV	TO220
7812	14.5–35 V	12 V @ 1 A	0.7%	0.2%	75 μV	TO220
7815	17.5–35 V	15 V @ 1 A	1%	0.055%	90 μV	TO220
7905	−7 to −30 V	−5 V @ 1 A	0.3%	0.3%	125 μV	TO220
7912	−14.5 to −35 V	−12 V @ 1 A	1%	0.07%	300 μV	TO220
7915	−17.5 to −35 V	−15 V @ 1 A	1%	0.055	375 μV	TO220

Line regulation is the percentage reduction of the output when the input changes. Load regulation is the percentage reduction of the output voltage as the output load changes.

24 Higher Electronics

Figure 1.5.3 *A negative voltage regulator*

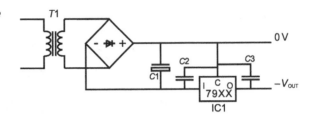

(1) The maximum voltage in must take into account the transformer regulation.
(2) $C1$ must be designed so that the rectified input supply does not dip below the minimum I_C input voltage.
(3) Suppression capacitors $C2$ and $C3$ must be mounted as close as possible to IC1.
(4) The output must have some form of loading. If it doesn't, there can be a high frequency (~5 MHz) oscillation superimposed on the d.c. output voltage. An easy solution might be to load the output with an LED indicator on the output.

Figure 1.5.3 shows the negative voltage equivalent.

1.5.1 Problems

(1) Design a 1 A, 5 V mains PSU (include selection of the mains transformer, rectifier and smoothing capacitor).
(2) Design a $+12$ V and -12 V 300 mA mains PSU.

If the output needs to be adjustable then it is possible to adapt the 78xx for positive voltages and the 79xx for negative voltages. Figure 1.5.4 shows a circuit which will allow this adjustment.

The IC will maintain the voltage between O and C at a constant 5 V. Thus, the base of Q1 is approximately 5.6 volts. In this case:

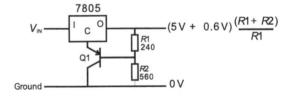

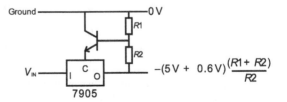

Figure 1.5.4 *Adjusting the three-terminal regulator*

Figure 1.5.5 *The LM317*

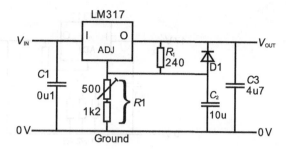

$$V_{REG} + 0.6 = V_o \frac{R2}{R1+R2}$$

$$\text{or} \quad 5.6 = V_o \frac{560}{560+240}$$

which gives $V_o = 8$ V.

Variable voltage three-terminal regulators

Of course, it is always better to use the correct tool for the job, and in this case the better tool might be something like an LM317 adjustable voltage regulator. This device does not have a common terminal.

Internally, it has a 1.25 V 'bandgap' reference which it maintains across 'OUT' and 'ADJ'. This reference is more accurate and reliable than that used in the 78xx/79xx regulators and so the output is more predictable.

Figure 1.5.5 shows an LM317 set up to provide a 9 V supply. C2 and D1 are not strictly necessary for correct operation. However, C2 improves the ripple performance, while D1 discharges C2 in the event of the output being short-circuited.

The minimum voltage which the LM317 can provide is 1.25 V, while the maximum is only limited by the fact that the input–output differential must not exceed 40 V. In its basic form, the LM317 will provide up to 1.5 A (with suitable heatsinking), but other derivative three-terminal adjustable regulators can provide 3 A (LM350), 5 A (LM338) or even 10 A (LM396). There are also variants which will work at higher input–output voltage differentials. The formula for calculating the output voltage is similar to that in the previous exercise, but adds in a term due to the current sourced by the I_{ADJ} terminal. Since I_{ADJ} is typically 100 µA, the error term of $I_{ADJ} \times R2$ may or may not be significant depending on the value of R2:

$$V_{OUT} = V_{REF}\left(\frac{R1+R2}{R1}\right) + I_{ADJ}R2$$

Problems 1.5.2

(1) Devise an LM317-based regulator to provide $+6$ V out from a 10 V source.
(2) With reference to Figure 1.5.6, devise a circuit which provides a switchable 5, 9, 12 V output from the 17 V input.

Figure 1.5.6

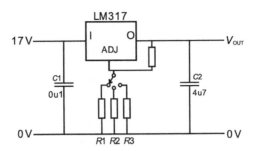

1.6 Heatsinks

When linear power ICs are used, they normally require cooling. The conventional method is to fix them to metal heatsinks which in turn can be cooled naturally or have air forced across them by fans. Every semiconductor will have a maximum rated junction temperature with which it can cope – usually in the range of 125–200 °C. If a long and healthy life is expected out of power devices, then it pays to keep them much cooler than this.

Every watt of power consumed will cause a rise in temperature. When a component is connected to a heatsink, a thermal circuit exists along which heat will flow from the silicon die to the surrounding air. The term 'thermal resistance' is used to describe the ease (or otherwise) with which heat will flow. Thermal resistance has the mathematical unit of °C per watt. Figure 1.6.1 shows four types of heatsink, together with their thermal resistances.

The quoted thermal resistance will be worsened if the device is not securely fixed to the heatsink. Special bonding pads (such as

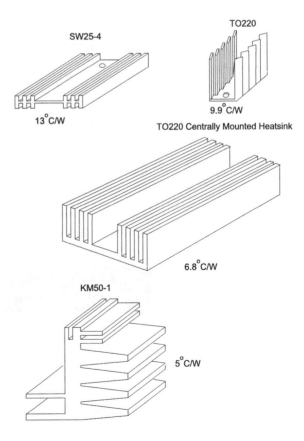

Figure 1.6.1 *Heatsinks*

Power supplies

'Sil-Pad' and 'Thermasil') are available which provide not only electrical insulation but also good thermal conduction. Fixing is usually by screw and nut, but some heatsinks clip on. The tab of the IC is usually connected electrically to one of the terminals, and sometimes although the heat sink pad is non-conductive, the fixing screw is. In these circumstances, either a nylon screw is used or an insulating insert passes through the mounting hole of the IC tab.

The human body perceives 70 °C as hot to the touch. If the ambient temperature inside an equipment is 30 °C, then the heatsink must only allow a 40 °C temperature rise. (That is if the heat sink is externally touchable. Even so, the remarks above should make it clear that it is unwise to allow the semiconductor temperature to get too hot.)

Example 1.6.1

What minimum rating of heatsink would be required for the circuit above?

The 7812 dissipates $(18-12)\,V \times 1\,A = 6\,W$. If heatsink temperature is not to rise more than 40 °C above ambient, a suitable heatsink should have a thermal resistance better than $40°C/6\,W = 6.6\,°C/W$.

Problems 1.6.1

(1) A 15 V_{RMS} mains transformer is full wave bridge rectified, smoothed with a 4700 μF capacitor and the resultant voltage used as the input to an LM317 setup to provide 10 V at up to 1.5 A:
  (a) Calculate the value of the other passive components needed.
  (b) Calculate the minimum value of heatsink thermal resistance.
  (c) Suggest a suitable heatsink.
(2) Using a 78M05 as the voltage regulation device, design a simple operational +5 V power supply unit. Your design should specify and cost (one-off) **all** the components which would be needed with the exception of pcb and case.
(3) Why should the fins of the heat sink be mounted vertically in normal use?

Electrolytic capacitors

An interesting thing about 'ordinary' electrolytic capacitors is that they have a life which is rated at only 2000 hours if operated

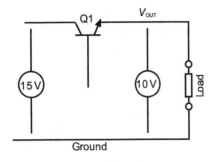

Figure 1.7.1 *Schematic of a simple current limit*

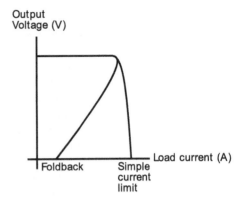

Figure 1.7.2 *Foldback action*

continuously at 85 °C. When designing your power supply, take care about component mounting positions and heat dissipation arrangements. If the internal temperature is found to be quite warm, consider using 'high-temperature' electrolytics.

1.7 Current limiting vs. foldback

There is a problem with the simple current limit scheme as shown, for example, in Figure 1.7.1.

If current limiting is set up to cut in when the load current exceeds 1 A, then up until that point Q1 will dissipate $(15\,V - 10\,V) \times 1\,A = 5\,W$. The heatsink will be designed to handle this (normal) rate of power dissipation. However, if the load is short-circuited, the output voltage is now 0 V, while the load is still drawing 1 A. The power dissipation is now $(15\,V - 0\,V) \times 1\,A = 15\,W$. Clearly there is a problem. To counteract this tendency, current foldback was developed.

The principle is shown in Figure 1.7.2. Beyond the cut-off point, not only is the output voltage limited, but the short-circuit output current is 'folded back' as well. The circuit of Figure 1.7.3 shows the modification to the basic current limit circuit which will create this foldback arrangement. One note of caution: if the foldback is set to too low a value, the circuit may have difficulty starting up into a high (but not excessive) load current. A good rule of thumb is to ensure that the folded-back current is never designed to be less than one-third of the maximum current.

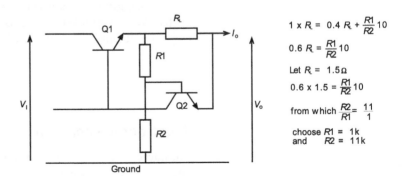

Figure 1.7.3 *A foldback circuit*

$1 \times R_L = 0.4\,R_L + \frac{R1}{R2}\,10$

$0.6\,R_L = \frac{R1}{R2}\,10$

Let $R_L = 1.5\,\Omega$

$0.6 \times 1.5 = \frac{R1}{R2}\,10$

from which $\frac{R2}{R1} = \frac{11}{1}$

choose $R1 = 1\,k$
and $R2 = 11\,k$

Power supplies

The potential divider of $R1$–$R2$ ensures that as the load is progressively increased, the maximum output current is increased (see box).

$$I_o \times R_L = \frac{R1 + R2}{R2} V_{BE(Q2)} + \frac{R1}{R2} V_o$$

The voltage drop across $R_L = V_E - V_o = I_o R_L$ ──(1)

When $V_B - V_o = V_{BE(Q2)}$ ──(2) then Q2 will turn ON.

V_B can be expressed in terms of V_E as

$$V_B = V_E \frac{R2}{R1 + R2}$$ ──(3) [Eliminating V_B between (1)(2)]

So:
$$V_E \frac{R2}{R1 + R2} - V_o = V_{BE(Q2)}$$ ──(4)

Eliminating V_E between (1) and (4) gives

$$I_o R_L = \frac{R1 + R2}{R2} V_{BE(Q2)} + \frac{R1}{R2} V_o$$

$V_o = 10\,V$; $I_o = 1\,A$. Let the short circuit current be $0.4\,A$. Then

$$I_o R_L = \frac{R1 + R2}{R2} V_{BE(Q2)} + \frac{R1}{R2} V_o$$

As current limiting commences: When the output is short circuited:

$$1 \times R_L = \frac{R1 + R2}{R2} 0.6 + \frac{R1}{R2} 10 \qquad 0.4 \times R_L = \frac{R1 + R2}{R2} 0.6$$

Substituting

Problems 1.7.1

(1) Design a foldback circuit which will allow a 12 V output at a maximum of 0.5 A, and yet foldback to 0.2 A when presented to a short circuit.

(2) Design a power supply unit (from transformer to output) which will provide a regulated 18 V at up to 0.8 A. Foldback should be designed in to limit the short-circuit current to 0.4 A.

(3) In the initial example, the ratio of $R2:R1$ was 11:1. Could I have used:
   (a) 110k and 10k;
   (b) 1.1M and 110k?

If not, why not?

1.8 Voltage multipliers

Voltage multipliers are simply made from diodes and capacitors. They convert an a.c. input signal into a larger d.c. voltage output. They are useful for building EHT (Extra High (voltage) Tension) supplies for systems which require a high voltage but low current source.

Before we look at the various types of multiplier, we must revise two useful diode-capacitor circuits, the understanding of which are fundamental to that which follows.

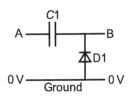

DC restoration

It is a fact that it is not possible to change instantly the voltage across a capacitor. There is a limit to the rate at which current can flow into, or out from, the capacitor plates.

When V_A goes positive, V_B follows it. When V_A goes negative, V_B follows it, but forward conduction through diode D1 limits the voltage at B to -0.6 V. When V_A steps positive by $2 \times V_{pk}$, V_B follows it, producing the waveforms shown.

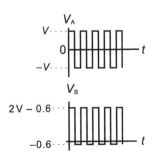

Peak clamp

In Figure 1.8.2, D2 half-wave rectifies the a.c. signal and the peak voltage (less the forward diode drop) is stored on the capacitor $C2$.

The slight discharge droop on the waveform is dependent on the load connected to B. If B is open circuit, or connected to a very high impedance load, the output will be smooth. If the load is lower impedance, then during the a.c. 'low' period, the capacitor voltage will droop as shown.

Figure 1.8.1 *A d.c. restoration circuit*

Supply doublers

In Figure 1.8.3, when X is more positive than Y, D1C1 act as a positive peak hold circuit. When X is more negative than Y, D2C2 act as a negative peak hold circuit.

For example:
If X–Y is 12 V_{RMS}:

$$C1 \text{ would store } + (12 \times \sqrt{2} - 0.6 \text{ V}) = +16.4 \text{ V}$$

$$C2 \text{ would store } - (12 \times \sqrt{2} - 0.6 \text{ V}) = -16.4 \text{ V}$$

That is:

$$V_o = 32.8 \text{ V}$$

What would be the output voltage of such a circuit using a 15 V transformer secondary?

A different approach resulted in the Cockroft–Walton multiplier of Figure 1.8.4. C1D1 form a d.c. restoring circuit while D2C2 form a peak clamp circuit. The output V_B will be $V_A(\text{pk}) - 0.6 - 0.6$.

The circuit can be extended indefinitely – but becomes less efficient as it does so (Figure 1.8.5).

The advantage of these circuits is that the input does not need to be a floating transformer secondary. It is ground referenced and can be easily driven from IC sources. Each capacitor need only be rated at a value in excess of the peak-peak input voltage. The series

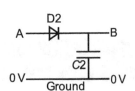

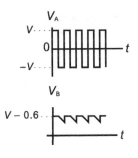

Figure 1.8.2 *A peak clamp*

Power supplies

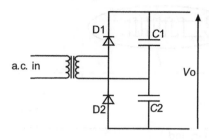

Figure 1.8.3 *A voltage doubler*

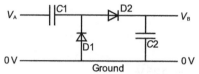

Figure 1.8.4 *An alternative voltage doubler*

nature of the capacitors produces very poor voltage regulation at the output as the load current increases.

An alternative approach is the parallel circuit (Figure 1.8.6):

- Disadvantage – all capacitors must withstand the full output voltage, which can be expensive for a large array.
- Advantage – better regulation can be achieved.

Voltage level converters

Voltage converters can be constructed from a simple 'diode pump' as in Figure 1.8.7.

C1D1 form a negative voltage d.c. restoration circuit, the peak value of which is clamped and stored by D2C2. Although it is not capable of generating significant amounts of current from the new negative supply at B, it is a useful technique for producing undemanding negative voltages. It is the basic principle behind several commercial ICs which generate their own negative supply.

Hickman multiplier

This is a development of the Cockroft–Walton multiplier, but uses fewer components. It is at its best when built with high capacity, low impedance (expensive) electrolytic capacitors. It does, however,

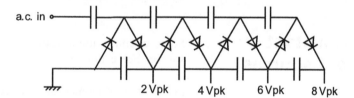

Figure 1.8.5 *A Cockroft–Walton voltage multiplier*

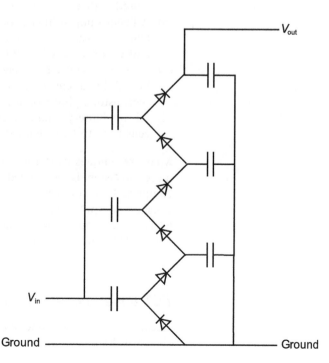

Figure 1.8.6 *An alternative Cockroft–Walton voltage multiplier*

Figure 1.8.7 *Diode pump*

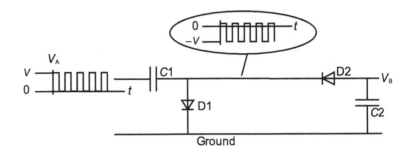

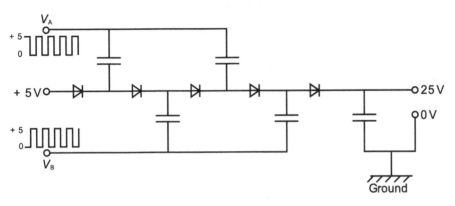

Figure 1.8.8 *The Hickman multiplier*

need a complementary square wave drive which is easily generated by integrated circuits.

1.9 Constant current sources

Some circuit configurations call for the use of a constant current source. For example:

(a) Voltage reference diodes provide a more stable output when biased with a constant current source.
(b) A linear ramp voltage is best generated by charging a capacitor from a constant current source. Since $I = C\,dV/dt$, keeping I and C constant will result in a linear change of voltage.
(c) Transducers which depend on a change of resistance can be biased by a constant current source. These would include thermistors, photoconductive cells, magneto-resistors, etc.
(d) NiCad battery charging circuits. Nickel–cadmium secondary cells need to be recharged at a constant current.

Many techniques exist for generating a constant current, and as more precision is demanded, such circuits get more and more complex. Like voltage references, they are expected to maintain a constant output regardless of loading, ageing or temperature effects. This section will concentrate on three techniques: using FETs, BJTs and op-amps.

FET-based constant current sources

Recall that the transconductance characteristic of an FET is as shown in Figure 1.9.1.

Power supplies

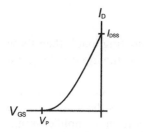

Figure 1.9.1 The FET constant current source

The internal constant current source is controlled by the gate-source voltage V_{GS}. Like all transistors, characteristic values alter within a wide range. For example, a 2N3819 n-channel JFET claims I_{DSS} in the range of 2 mA to 20 mA and V_P in the range -4 V to -8 V.

By connecting the gate directly to the source, the saturated drain-source current I_{DSS} will flow – i.e. somewhere between 2 mA and 20 mA.

The disadvantage of this circuit is that the value of I cannot be set. It is dependent totally on the I_{DSS} value of the JFET. A simple adjustment will allow the user to set I_{SINK} or I_{SOURCE} to any value below I_{DSS}.

As I_D passes through R_S, the voltage drop V_{GS} can be adjusted to modify the value of I_D.

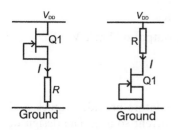

Figure 1.9.2 A current sink

BJT-based constant current circuits

Npn transistors tend to be used for current sinks and pnp transistors for current sources. They both work on the same principle – that of maintaining a constant voltage across a fixed resistor R_E. In each case, $R1$ biases the diode(s) and provides base current for the transistor (Figure 1.9.4).

V_{BE} is approximately 0.6 V while V_B is at 1.8 V ($3 \times$ diode forward voltage drops), I_C @ $1.2\,V/R_E$.

To set this to (say) 5 mA with a ZTX 300 ($h_{FE(MIN)}) = 100$:

$$R_E = 1.2\,V/5\,mA = 240\,W$$

$R1$ must deliver D1D2D3 bias current together with I_B. If $I_{DIODE} = 1$ mA and $I_B = 50\,\mu A$ then $R1$ is $(V_S - 1.8\,V)/1.05\,mA$.

Here lies the disadvantage of this circuit: the base voltage V_B depends on V_S. As V_S increases so will V_B.

Here is another CCS. Recall from Figure 1.9.5 that Z1 has minimal temperature sensitivity when selected as $\sim$5.6 V. However, its anode slope resistance still means that V_B is dependent on V_S. A better solution might be the circuit of Figure 1.9.6.

Temperature change is still a problem though.

The FET with gate coupled to the source provides a relatively constant current into the Zener diode of I_{DSS}. This makes the Zener more immune to supply voltage changes.

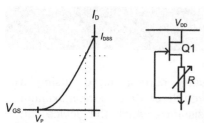

Figure 1.9.3 Adjustment of the current source

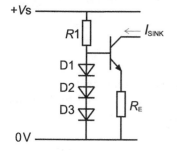

Figure 1.9.4 A BJT constant current source

Problems 1.9.1

(1) Design a pnp transistor-based 100 mA current source.
(2) Design an npn transistor-based 25 mA current sink.
(3) Use the circuit of Figure 1.9.6 as the basis of a 250 mA current sink.

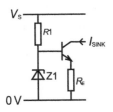

Figure 1.9.5

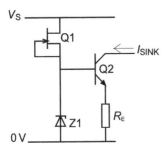

Figure 1.9.6 *An improved constant current source*

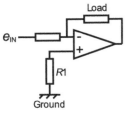

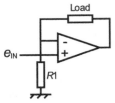

Figure 1.9.7 *Simple floating load current supplies*

Op amp solutions

As long as the circuit is provided with a 'floating load', then these are quite simple. The current is determined by the input voltage e_{IN} and the input resistor $R1$.

In each case in Figure 1.9.7, $i_{LOAD} = e_{IN}/R1$. A constant voltage at the input maintains a constant voltage across the load from the output to the virtual earth point at the operational amplifier input.

The circuit becomes more complicated when an earthed load is required. In this case, the Howland current source is the classical solution (Figure 1.9.8).

This relies on matched resistors for best performance, but when an analysis is carried out:

$$I_o = \frac{e_{IN}}{R5} \times \frac{R2}{R1}$$

With the values shown, this gives an output of 1 mA/V.

IC voltage regulator

The 78xx and 79xx families of IC regulators provide a quick and easy method of building a precision current source. The design of these devices is to maintain a fixed voltage between O and C. Figure 1.9.9 shows a 100 mA 5 V regulator. This would be changed into a 5 mA current source by connecting as shown in Figure 1.9.10.

In this case $5\,V/1\,k\Omega = 5\,mA$.

Problems 1.9.2

(1) Design a 0.5 A constant current source using a 7812 12 V, 1 A voltage regulator. What is the maximum and minimum load resistance which the device can supply?

(2) Use a 79L05 voltage regulator to perform as a 50 mA current sink.

(3) Design a 500 mA current source with the aid of a 7815 three-terminal voltage regulator. When operating from a 25 V unregulated supply, how much power will it dissipate? What rating of heatsink would be appropriate?

1.10 Switched mode power supplies

The series pass transistor which has been common to all of the linear power supplies so far studied is often expected to dissipate large amounts of power. For example, a 1 A 5 V supply which is fed from a 10 V d.c. input is only 50% efficient:

$$P_{in} = 10\,V \times 1\,A = 10\,W \quad P_{out} = 5\,V \times 1\,A = 5\,W$$

If the input voltage rises to 15 V:

Power supplies

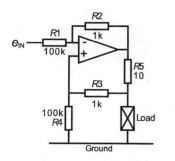

Figure 1.9.8 *The Howland current source*

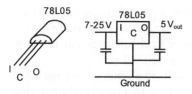

Figure 1.9.9 *A three-terminal voltage regulator as a voltage source*

o/p power delivered: $5\,\text{V} \times 1\,\text{A} = 5\,\text{W}$

i/p power consumed: $15\,\text{V} \times 1\,\text{A} = 15\,\text{W}$

Hence the efficiency falls to 33%.

In this case, the 10 W dissipated by the pass transistor would mean fitting it to a substantial heatsink. This inefficiency is the main disadvantage of a linear supply.

A transistor turned fully on has a very low voltage drop (<0.1 V for a BJT). This property is well known and used in device drivers and electronic switches. Switching is electrically noisy and so most SMPSs operate at a fixed frequency so that filtering is easier to perform. Control of the power transferred by the switching action is usually by pulse width modulation techniques. This defines the on period as a proportion of the total period.

The mark–space ratio or duty-cycle d is given by:

$$d = \frac{t_{ON}}{t_{ON} + t_{OFF}}$$

Figure 1.10.1 shows two waveforms with:

(i) 10% mark–space ratio; and
(ii) 75% mark–space ratio.

The longer the ON period, the more time is available for energy to be transferred by the switch. Hence the average voltage is higher.

There are two features which identify switch mode power supplies:

(1) transistors running efficiently as switches;
(2) inductors used as energy storage elements.

Table 1.10.1 shows a comparison of the properties of the two power supply technologies.

There are **four** main types of SMPS:

- Forward (or buck) converter — Step down
- Boost regulator converter — Step up
- Flyback (or buck-boost) converter — Voltage inversion/step down/step up
- Cuk converter — Voltage inversion/step down/step up

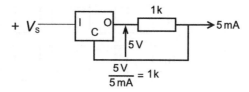

Figure 1.9.10 *A three-terminal regulator connected as a current source*

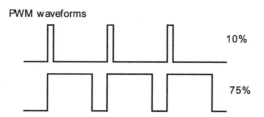

Figure 1.10.1

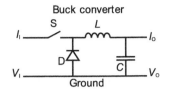

Figure 1.10.2 *The buck converter*

Table 1.10.1

Parameter	Linear PSU	SMPS
Efficiency	Typically 30%	Typically 75–95%
V_i–V_o differential	Small	High
Noise	Low	High
P_{out}/volume	Low	High (up to 100 mV pk-pk)
Output ripple	Low (5 mV pk-pk)	High (50 mV pk-pk)

Buck converter

When switch S is closed, current flows through the inductor to the reservoir capacitor and to the load. A magnetic field builds up in the inductor.

When the switch opens, a back emf is created in the inductor and the end nearest diode D tries to go to a negative potential. Diode D is forward biased, and as the magnetic field collapses, current flows through the inductor and onto the capacitor/load.

The buck configuration always results in an output voltage which is lower than the input voltage. There are two modes of operation, dependent on whether the inductor current falls to zero or not. Control is simpler (if slightly less efficient) in the continuous mode.

Figure 1.10.3 shows typical waveforms which can be observed in the circuit. V_{OUT} is set by the duty-cycle of the waveform, and the voltages across the switch and diode are shown at (i) and (ii). In the steady state, I_o and L are fixed and according to the relationship $V = L\,dI/dt$, $V \times dt$ represents a constant value. Hence the areas A and B in figure (iii) must necessarily be of the same area, representing as they do the product of voltage and time. Electronic control is used to vary the duty cycle of the switch to maintain a constant output voltage. The inductor must be chosen:

(i) so that the core does not magnetically saturate;
(ii) so that sufficient energy can be stored for the application.

The diode must be a fast-acting variety such as a Schottky diode.

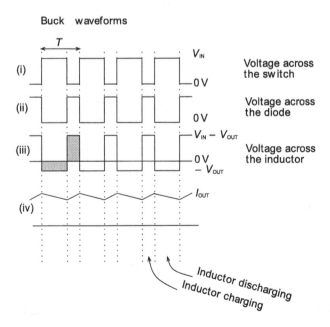

Figure 1.10.3 *Buck converter waveforms*

With the switch closed:
$$V = L\frac{dI}{dt}$$
$$V \times dt = L \times dI$$
$$(V_{IN} - V_{OUT})t_{ON} = I\,dI$$

With the switch open:
$$-V_{OUT}t_{OFF} = -L\,dI$$

from which:
$$-V_{OUT}t_{OFF} = -L\,dI$$
$$(V_{IN} - V_{OUT})t_{ON} = V_{OUT}t_{OFF}$$

and hence
$$V_{OUT} = V_{IN}\frac{t_{ON}}{t_{ON} + t_{OFF}}$$

Problems 1.10.1

(1) (a) What duty cycle would be needed to drop a 9 V supply down to 5 V?
   (b) If the switching frequency is 50 kHz, what would be the main switch ON and OFF times?
(2) Why is it important that the magnetic core never saturates?
(3) If a 12 V input has a switch ON period of 0.02 ms and OFF period of 10 µs:
   (a) What would be the output voltage?
   (b) What is the switching frequency?

Figure 1.10.4 shows the circuit schematic of component blocks which would be needed to create a buck regulator.

While it is perfectly possible to design discrete component SMPSs the trend is to use integrated solutions.

Figure 1.10.5 shows the Linear Technology LT1074, a 5 pin TO220 package. It operates at a fixed frequency of 100 kHz and is capable of supplying up to 6.5 A at an output voltage from 2.5 V to 50 V.

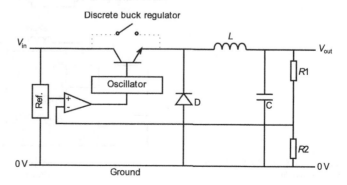

Figure 1.10.4 *Buck regulator schematic*

Figure 1.10.5 *An IC SMPS*

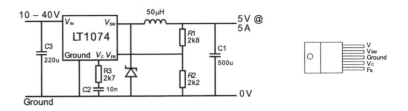

The boost regulator

The boost regulator can produce an output which is greater than the input, or of opposite polarity to the input. In Figure 1.10.6, diode and capacitor *C* are a form of peak clamp circuit as seen in Figure 1.8.3. Inductor *L* and switch S act in the same way as the ignition circuit on a car. When switch S is closed, current flows through *L* to ground. When switch S is opened, a voltage is developed across *L* according to:

$$V = L\frac{\mathrm{d}i}{\mathrm{d}t}$$

The on–off ratio of the switch controls the amount of current which builds up in the inductor and hence controls the output voltage.

Figure 1.10.7 shows a useful commercial circuit from MAXIM which will produce a regulated +5 V, 240 mA output from a 1.5 V input. This circuit could be powered by a single alkaline cell.

The buck–boost regulator

A buck–boost regulator can develop an output less than or greater than the input voltage.

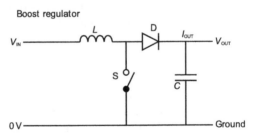

Figure 1.10.6 *A boost regulator*

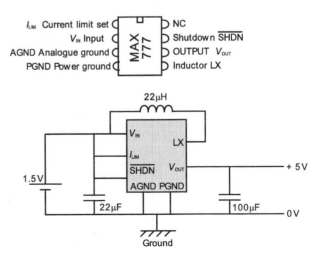

Figure 1.10.7 *A step-up switched mode power supply*

Power supplies

Figure 1.10.8 A buck–boost circuit

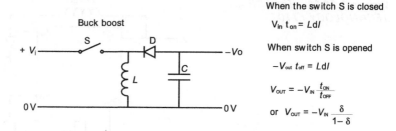

When the switch S is closed

$V_{in} \, t_{on} = L \, dI$

When switch S is opened

$-V_{out} \, t_{off} = L \, dI$

$V_{OUT} = -V_{IN} \dfrac{t_{ON}}{t_{OFF}}$

or $V_{OUT} = -V_{IN} \dfrac{\delta}{1-\delta}$

In an application circuit, the switch S would be controlled in the usual way by monitoring the output voltage and comparing it with a fixed reference. The difference between these two voltages would control the mark–space ratio of the ON–OFF of the switch.

The flyback regulator

This might seem somewhat more familiar in that it has a transformer which is converting the switched a.c. at the primary into a voltage at the secondary.

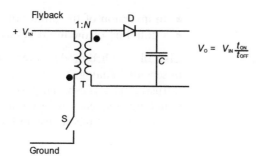

Figure 1.10.9 A flyback regulator

Control is achieved by measuring the output voltage and using the difference between this and the reference to adjust the switching ratio of the electronic switch S.

More examples

Figures 1.10.10 and 1.10.11 show two more common SMPSs which are popular in many applications.

Figure 1.10.10 shows a high efficiency (>90% claimed) step-down converter which can provide from 10 mA to 1.5 A. The input supply is restricted to 3.5–16.5 V, and in the configuration shown requires an external MOSFET to boost the output current.

Figure 1.10.11 shows a versatile monolithic high power switching regulator which can be operated in all of the standard

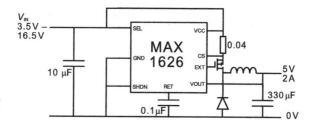

Figure 1.10.10 The MAXIM MAX 1626

Figure 1.10.11 *The LINEAR TECHNOLOGY LT1070*

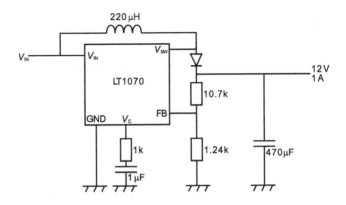

configurations. It can output a maximum of 5 A and work from supplies from 3 V to 60 V.

The charge pump

The charge pump is a circuit which operates using charge storage in capacitors. It does not use inductors at all (Figure 1.10.12).

There are two outputs which can be used to produce:

- an inversion of the input voltage (−5 V out for +5 V in);
- a doubling of the input voltage (+10 V out for +5 V in).

Like all switched mode power supplies, they must be used with care in sensitive analogue circuits, since the switching frequency can be intrusive. This is especially so with the ICL7660, since the nominal switching frequency is in the audio band at 10 kHz. Nevertheless, like all tools, it has its use in the right place.

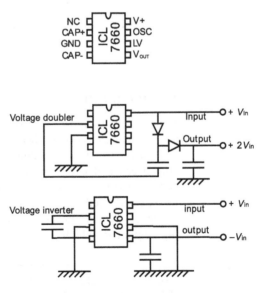

Figure 1.10.12 *The charge pump*

2 Feedback (1)

Summary

In this chapter, the first of the two feedback chapters, you will learn about the basic concepts of feedback: what it is and why it is used. It was introduced at a basic level in Chapter 1 (Power supplies).

2.1 Introduction

Feedback is a technique where a proportion of the output of an amplifier is fed back and recombined with the input. There are two types of feedback:

- positive feedback;
- negative feedback.

Figure 2.1.1 shows the phase relationships between the input signal and the fed-back signal for positive and negative feedback:

(i) If the signal fed back is in phase (0°) with the input, the feedback is called positive. When it is in phase, the signals add up to make a bigger signal. The output is increased in size. This usually leads to instability, as will shortly be explained.

(ii) If the signal fed back is out of phase (180°) with the input, the feedback is called negative. When it is out of phase, the signals subtract to make a smaller signal.

Schematically the feedback loop can be shown as in Figure 2.1.2. It is drawn as a voltage-in, voltage-out amplifier in this circuit, but we shall shortly meet other combinations where the input and/or the output could be current.

A represents the gain of the amplifier:

$$A = V_o/V_e \quad \text{or rearranged} \quad V_o = A \times V_e$$

β represents the proportion by which the output is reduced before it is combined with the input:

42 Higher Electronics

Figure 2.1.1 *Phase relationships*

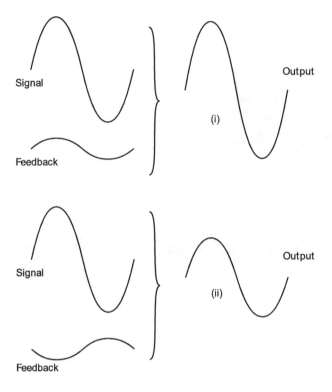

$$\beta = V_f/V_o \quad \text{or rearranged} \quad V_f = \beta \times V_o$$

The input V_i is combined with the feedback voltage V_f by the mixer. This circuit schematic is drawn in two ways in Figure 2.1.3: one to represent positive feedback and one to represent negative feedback.

Basic equations (+ve)

$$V_o = AV_e$$
$$V_f = \beta V_o$$
$$V_i + V_f = V_e$$

Now, combining these:

$$V_i + \beta V_o = \frac{V_o}{A}$$
$$V_i = \frac{V_o}{A} - \beta V_o$$
$$V_i = V_o\left(\frac{1}{A} - \beta\right)$$
$$V_i = V_o\left(\frac{1 - A\beta}{A}\right)$$
$$\frac{V_o}{V_i} = \frac{A}{1 - A\beta}$$

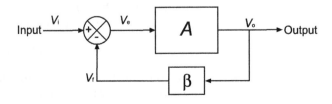

Figure 2.1.2 *A general feedback amplifier*

Figure 2.1.3 *Positive and negative feedback*

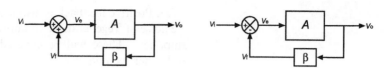

Basic equation (−ve)

$$V_o = AV_e \qquad V_f = \beta V_o \qquad V_i - V_f = V_e$$

Now, combining these:

$$V_i + -\beta V_o = \frac{V_o}{A}$$

$$V_i = \frac{V_o}{A} + \beta V_o$$

$$V_i = V_o\left(\frac{1}{A} + \beta\right)$$

$$V_i = V_o\left(\frac{1 + A\beta}{A}\right)$$

$$\frac{V_o}{V_i} = \frac{A}{1 + A\beta}$$

Positive feedback results in a denominator term of and negative feedback a denominator term of $1 + A\beta$. So how does this affect amplifier performance?

2.2 Positive feedback

When $A \times \beta = 1$ in the expression:

$$\frac{V_o}{V_i} = \frac{A}{1 - A\beta}$$

then the denominator reduces to 0. Divide any number by zero and the result is an infinitely large number. In practice, the circuit oscillates. Imagine a circuit with a gain of 10 and a feedback factor of 1/10. When a 1 volt a.c. signal is input, the output becomes 10 V. The feedback factor reduces this to 1 V to be added to the input. At this point, the input could be removed, and the amplifier would, in theory, continue oscillating. If the 1 V input is still present, the output is increased still further until it is limited by the amplifier supply voltage rails. (A linear amplifier cannot produce an output more positive or more negative than its power supply voltages.) In Chapter 8 (Oscillators), we shall see how this property is used to make circuits oscillate at one particular frequency.

Positive feedback can often be demonstrated in audio systems which use a microphone, amplifier and loudspeaker. A 'howl-around' is created when the microphone picks up the output of the loudspeaker and amplifies it. The microphone picks up this louder sound and the amplifier amplifies it. The microphone picks up ... and so on. It is not considered one of the more useful applications of positive feedback.

Positive feedback is used more beneficially in digital (on–off) switching circuits to improve the speed of response and immunity to noise. Positive feedback is also used in radio frequency amplifiers in the guise of 'regeneration' to improve the gain performance of those amplifiers. These will be covered in Chapter 9 (Radio frequency and other techniques).

2.3 Negative feedback

The rest of this chapter will be concerned with the use and application of negative feedback.

The formula for the closed loop gain of a negative feedback amplifier has been given as:

$$\frac{V_o}{V_i} = \frac{A}{1 + A\beta}$$

Problems 2.3.1

(1) If $A = 1000$ and $\beta = 1/25$, calculate the closed loop gain (V_o/V_i) of the amplifier.
(2) If $A = 10\,000$ and $\beta = 1/25$, calculate the closed loop gain of the amplifier.

In these problems, A, the open loop gain increases tenfold, yet the overall closed loop gain only increases by a factor of 2.2%.

To simplify the formula: if A is very large, $A\beta \gg 1$, so the formula approximates to:

$$\frac{V_o}{V_i} = \frac{A}{A\beta} = \frac{1}{\beta}$$

Problems 2.3.2

(1) If an amplifier has a gain of 10 000, what is the approximate closed loop gain when a feedback factor of 1/25 is applied to it?
(2) A voltage of 10 mV is applied to an amplifier with an open loop gain of 1500 and a feedback factor of 0.2. What is the approximate output voltage?
(3) What feedback factor is required to produce an overall closed loop gain of 50 when the open loop gain is 1250?
(4) An amplifier has a forward gain of 58 000. If feedback is derived from an output circuit as shown in Figure 2.3.1, what is the closed loop gain?

As a hint, work out the feedback voltage as a proportion of the output voltage and you will have the value of β.

(5) An amplifier has a forward gain of 80 000. Its output is fed into the network of Figure 2.3.2. VR1 allows adjustment of the gain.

Calculate the values of β when the potentiometer:

(a) is at end X of VR1;
(b) is at end Y of VR1.

Hence calculate the maximum and minimum gain adjustment which can be made by varying VR1.

In general, provided that A is large, the voltage gain of an amplifier is approximately the inverse of the feedback factor. So what is 'large' in amplifier terms?

If we were to look at the typical gain of a single small signal amplifier, we would get values as shown in Table 2.3.1. (A small signal amplifier is a simple voltage or current amplifier, not one which provides significant power amplification into a load.)

Table 2.3.1

Technology	Voltage gain
Bipolar junction transistors (BJT)	100–200
Field effect transistors (FET)	5–20
Operational amplifiers (op amps)	100 000–10 000 000

The formula: $|\text{gain}| \approx 1/\beta$ applies when $A \times \beta \gg 1$. Thus, it depends which technology is being used as to whether $1/\beta$ is a close approximation to closed loop gain.

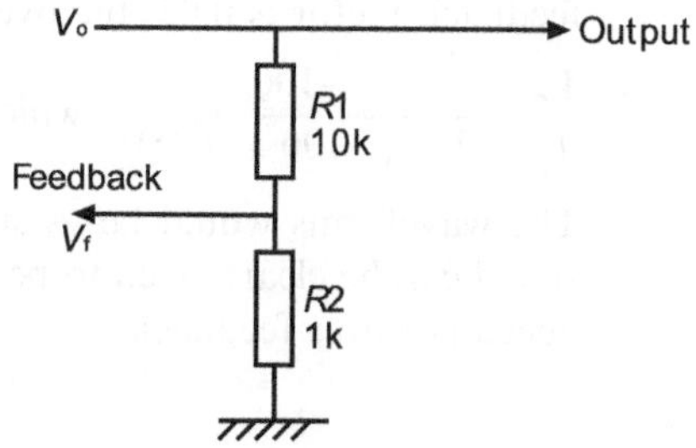

Figure 2.3.1

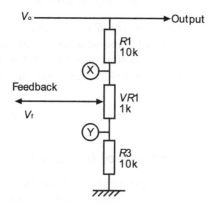

Figure 2.3.2

46 Higher Electronics

Figure 2.3.3 *Negative feedback with an inverting amplifier*

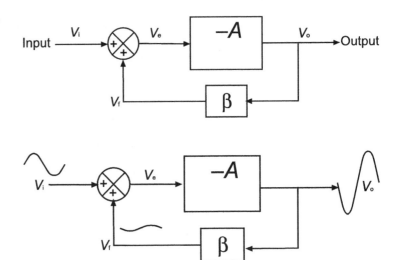

Figure 2.3.4 *Waveform phases*

The simple schematic of the right-hand diagram of Figure 2.1.3 does not cover all the possibilities of negative feedback. Sometimes, the 'mixer' circuit does not perform quite as shown. Figure 2.3.3 shows another circuit schematic which demonstrates negative feedback.

In this instance, the amplifier inverts, i.e. the output is out of phase with the input. This means that the feedback voltage V_f is out of phase with the input voltage V_i, so by definition, the feedback is negative.

The formula for feedback still applies in that:

$$\frac{V_i}{V_o} = \frac{A}{1 - A\beta}$$

Since the mixer used has two positive inputs, the formula should be as shown. For example, if the amplifier gain is -100 and the feedback factor is 0.05, the overall gain would be:

$$\frac{V_o}{V_i} = \frac{-100}{1 - (-100 \times 0.05)} \quad \text{which simplifies to} \quad \frac{-100}{6} \quad \text{i.e.} \quad -16.7$$

The waveforms would be as shown in Figure 2.3.4. The feedback signal can be clearly seen to be of the opposite phase to the input – hence negative feedback.

2.4 Benefits of negative feedback

There are many benefits to negative feedback, including:

(i) increase of bandwidth;
(ii) reduction of amplifier noise;
(iii) reduction of amplifier distortion;
(iv) modification of input and output impedances.

We shall now expand on each of these.

Increase of bandwidth

All linear amplifiers have the property of constant gain–bandwidth product (Constant GBW). That is, if you reduce the gain, you

Figure 2.4.1 *Gain–bandwidth product graph*

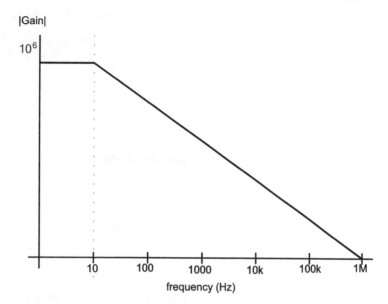

increase the bandwidth. Gain–bandwidth product has the unit of Hertz.

Figure 2.4.1 shows a typical gain–bandwidth product. These graphs usually have axes of frequency (in logarithms) against the modulus of gain (|Gain| in terms of amplitude only – no reference to phase). It is typical of a 741-type integrated circuit Operational Amplifier. The GBW is typically 10^6 Hz. Open loop, the bandwidth would only be 10 Hz. If negative feedback was applied so that $\beta = 1/100$, the gain would be reduced to ~100 and the bandwidth would be 10 kHz. (Gain × bandwidth = GBW or $100 \times 10\,000 = 10^6$.)

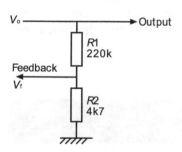

Figure 2.4.2

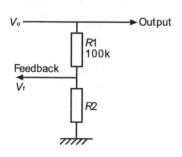

Figure 2.4.3

Problems 2.4.1

(1) An integrated circuit operational amplifier based amplifier has an open loop gain of 10^5 and a bandwidth of 15 Hz:
  (a) What is the GBW?
  (b) What would be the bandwidth if 2% negative feedback was applied?
  (c) What would be the maximum gain expected if a bandwidth of 10 kHz was required?

(2) An amplifier has an open loop gain of 500 000 and an open loop bandwidth of 5 Hz:
  (a) Calculate the gain bandwidth product.
  (b) If the circuit of Figure 2.4.2 was used to derive the feedback, what would be the closed loop gain?

(3) If an amplifier has an open loop gain of 750 000 and its output is connected to the circuit of Figure 2.4.3:
  (a) What value of R2 would be required if the closed loop gain is to be 25?
  (b) If the GBW product is 1 500 000 Hz, what now

would be the bandwidth?

(4) An amplifier shows a closed loop gain of 30 with a bandwidth of 70 kHz. If the feedback factor is 0.04:
   (a) What is the open loop gain?
   (b) What is the open loop bandwidth?

Cascaded stages

To increase gain, amplifier stages have to be cascaded. So, for example, an amplifier with a voltage gain of 100 feeding into an amplifier of voltage gain 10 would amplify an input signal by $10 \times 100 = 1000$ times (or, in decibels, $20\,\text{dB} + 40\,\text{dB} = 60\,\text{dB}$).

$$\text{Voltage gain} = 10 = 20\,\text{dB}$$

$$\text{Voltage gain} = 100 = 40\,\text{dB}$$

Cascaded logarithms **add** while gain values **multiply**.

Example 2.4.1

A certain amplifier has a gain–bandwidth product of 105 Hz. If a 50 kHz bandwidth is required to be maintained, then its maximum gain would be 2.

A particular application needs a gain of 8 with an overall bandwidth of 50 kHz, so **three** stages would be needed ($2 \times 2 \times 2$).

If a gain of 10 is required, then **four** stages would be required. One of the stages could have a gain of 1.25 so that $2 \times 2 \times 2 \times 1.25 = 10$. The stages with a gain of 2 would have the required bandwidth of 50 kHz, while the stage with a gain of 1.25 would have a bandwidth of 80 kHz. When several stages are cascaded together, the overall bandwidth is set by the lowest common denominator: the bandwidth would be 50 kHz as in Figure 2.4.4.

Problems 2.4.2

(1) A certain integrated circuit operational amplifier has a gain–bandwidth product of 2×10^5 Hz. It is to be used to produce an amplifier with an overall |gain| of 100 while maintaining a bandwidth of 20 kHz. How many stages would be required?

(2) A microphone with 5 mV output is to have its signal boosted to 1 V whilst maintaining a 12 kHz bandwidth. Amplifiers with a GBW of 600 000 are available. How many stages of such amplifiers are required?

Figure 2.4.4

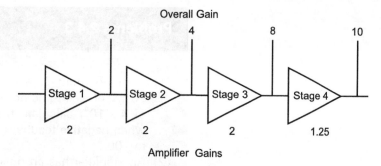

Reduction of amplifier noise

All electronic devices generate noise. In Chapter 6 (Noise), we will explore the types of noise, the mechanisms by which they arise and how to minimize noise in practical circuits. For now, let us assume that noise is an additional signal which is generated by the amplifier. It is completely unrelated to the input signal. Figure 2.4.5 shows a block diagram of a noisy amplifier.

The basic equations are:

$$V_o = A \times V_e + V_n \quad V_f = \beta V_o \quad V_e = V_i - V_f$$

So $V_o = A \times V_i - A\beta V_o + V_n$ which rearranges to:

$$V_o(1 + A\beta) = AV_i + V_n$$

$$V_o = V_i \frac{A}{1 + A\beta} + \frac{V_n}{1 + A\beta}$$

In this system, the amplifier has an extra term. It increases its input by A times, but adds the noise voltage V_n. Hence the term $V_o = AV_e + V_n$.

The open loop gain is reduced by the factor $(1 + A\beta)$. This is the same as previously. The noise, however, is also reduced by the factor $(1 + A\beta)$.

Example 2.4.2

An amplifier with open loop gain 1000 contributes 50 mV of noise to an input signal. If 3% overall negative feedback is applied, calculate the value of the closed loop gain and the noise voltage output.

From the above relationship:

$$V_o = \frac{1000 \times V_i}{1 + 33} + \frac{50\,\text{mV}}{1 + 33}$$

$$= 29.4 \times V_i + 1.47\,\text{mV}$$

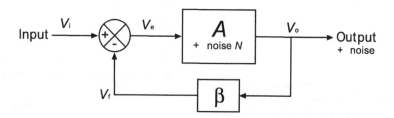

Figure 2.4.5

Problems 2.4.3

(1) With no input signal, a certain amplifier produces 70 mV of electronic noise. If it has an open loop gain of 4×10^4, calculate the reduction in noise voltage when negative feedback reduces the closed loop gain to 100.

(2) An amplifier has an open loop gain of 15 000. When the feedback factor is 1/25, the output noise voltage is 50 µV. What would the noise voltage be without feedback?

Reduction of distortion

Like noise, distortion is a corruption of the input signal caused by the amplifier itself. The most common form of distortion is harmonic distortion. For example, if a 1 kHz signal is input into an amplifier with a significant harmonic distortion figure, the output would contain frequencies as shown in Table 2.4.1.

Table 2.4.1

	Input		Output		
Component	signal	fundamental	2nd harmonic	3rd harmonic	4th harmonic
Frequency	1 kHz	1 kHz	2 kHz	3 kHz	4 kHz
	f	f	f_2	f_3	f_4
			$2 \times f$	$3 \times f$	$4 \times f$

In general, the higher the harmonic order of the distortion, the smaller the amplitude of the signal (Figure 2.4.6).

Total harmonic distortion (THD) is given by:

$$\text{THD} = \sqrt{V_2^2 + V_3^2 + V_4^2 + V_5^2 + \ldots}$$

(see Figure 2.4.7).

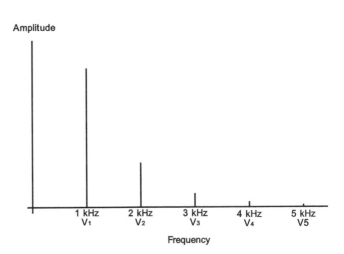

Figure 2.4.6 *The frequency spectrum of the output of an amplifier with harmonic distortion*

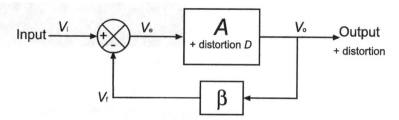

Figure 2.4.7 Block diagram of a feedback amplifier with distortion

Mathematics in action

When the amplifier introduces distortion, the signal fed back would be:

$$V_f = \beta V_o$$

$$V_f = \beta(V_{1o} \sin \omega T + V_{2o} \sin 2\omega T + V_{3o} \sin 3\omega T + V_{4o} \sin 4\omega T + \cdots)$$

where:

V_{1o} is the fundamental peak amplitude

V_{2o} is the second harmonic peak amplitude

V_{3o} is the third harmonic peak amplitude

etc.

The voltage into the amplifier after the mixer is given by $V_e = V_i - V_f$. As written, this equation is usually taken to represent RMS values, but if we account for the fact that the input is in reality a sine wave, then this can be presented as:

$$V_e = V_i \sin \omega T - \beta(V_{1o} \sin \omega T + V_{2o} \sin 2\omega T + V_{3o} \sin 3\omega T + V_{4o} \sin 4\omega T + \cdots)$$

Originally, the output was explained as $V_o = AV_e$, so the output voltage could be expressed as:

$$V_o = AV_i \sin \omega T - A\beta V_{1o} \sin \omega T - A\beta V_{2o} \sin 2\omega T - A\beta V_{3o} \sin 3\omega T - \cdots$$

where $V_o = V_{1o} \sin \omega T + V_{2o} \sin 2\omega T + V_{3o} \sin 3\omega T + \cdots$

Collating these two equations gives:

$$AV_i \sin \omega T = (1 + A\beta)V_{1o} \sin \omega T + (1 + A\beta)V_{2o} \sin 2\omega T + (1 + A\beta)V_{3o} \sin 3\omega T + \cdots$$

Note that each output term includes the expression $(1 + A\beta)$. In a 'normal' derivation, i.e. without distortion being accounted for, this is the factor by which the gain of the distortion term is reduced. Without feedback, the output would be:

$$V_o = AV_i = AV_{1o} \sin \omega T + AV_{2o} \sin 2\omega T + AV_{3o} \sin 3\omega T + \cdots$$

Hence, each harmonic term is reduced by the factor of $(1 + A\beta)$.

Problems 2.4.4

(1) Without feedback, a high gain amplifier introduces 10% total harmonic distortion. How much will this distortion be reduced to when the overall amplifier (closed loop) gain is reduced to 25?

(2) An amplifier with open loop gain of −10000 has a total harmonic distortion of 2%. If 3.3% negative feedback is applied, and the amplifier is presented with a 100 mV signal at its input, what would be:
   (a) the signal output voltage;
   (b) the distortion output voltage?

(3) A microphone signal needs to be boosted from 15 mV to 1 V before being input to a power amplifier. The signal amplifier being used to do this has a gain of 2×10^5, but introduces 3% harmonic distortion:
   (a) What feedback factor is required?
   (b) What will be the bandwidth of the amplified signal?
   (c) What will be the amplitude of the distortion?

(4) An amplifier with an open loop gain of 1 000 000 introduces 100 μV of noise and 5% total harmonic distortion. What would be the output noise and distortion when:

   (a) 2% negative feedback
   (b) 20% negative feedback

   is applied?

2.5 Electronic mixer circuits

The mixer has, so far, been given the system symbols as shown in Figure 2.5.1. Figure 2.5.1(i) has been used more often in the circuits so far, and can be electronically achievable in at least three different ways:

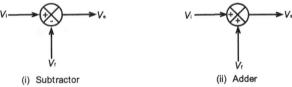

Figure 2.5.1 *(i) a subtractor; (ii) an adder*

(i) the long tailed pair;
(ii) the base–emitter comparator;
(iii) the comparator.

The long tailed pair

In Figure 2.5.2, R1 and R2 are the output load resistors while Q1 and Q2 are the comparison transistors. The way that they have their emitters connected together gives rise to the term 'long tailed pair'. For best performance:

(a) Q1 and Q2 should be h_{FE} matched;
(b) I_E should ideally be derived from a constant current source, but the circuit is occasionally used with a simple resistor in the tail.

Matched transistors can be bought in packages such as the BCY87 (Figure 2.5.3). Constant current sources were discussed in Chapter 1 (see page 32).

The circuit also works in its pnp version as in Figure 2.5.4.

Full analysis of this circuit can be complex, since as with any difference amplifier, the analysis is split into two sections:

(a) the differential mode analysis – how the circuit amplifies the voltage difference between the two inputs;
(b) the common mode analysis – how well the amplifier ignores the same signal when it is applied to the two inputs.

These two values give rise to a figure of merit for differential amplifiers called the common mode rejection ratio (CMRR):

$$\text{CMRR} = (\text{differential gain})/(\text{common mode gain})$$

Its value is usually quoted in decibels.

A simplified analysis of the long tailed pair follows. To work through it, you should be aware of the small signal analysis model of the BJT when operating in a common emitter mode (Figure 2.5.5).

This model is a quick and easy way of working out the a.c. performance of simple transistor circuits. There are more complex models which can be optimized for d.c., a.c./d.c. at low medium or high frequency operation. In general, the more complex the model, the more accurate can be the analysis. Semiconductor manufacturers provide models for new and existing devices for incorporation into software analysis tools such as SPICE, ECA, EDA, Workbench, etc. These are all commercial packages which can be used to analyse the d.c. and a.c. performance of electronic circuits.

The model shown is one of the simplest, and gives only a rudimentary insight into the expected performance. However, for a first attempt at circuit parameters, it is still quite useful. It only works for a.c. signals and, as you can see, it only tries to model the dynamic a.c. input resistance r_i and the current amplifier. This has the symbol β or h_{fe} and represents:

$$\beta = h_{FE} = \frac{i_C}{i_B}$$

The current amplifier produces an output i_C which in an ideal transistor would be an amplification of the input current i_B.

Another common convention is to work in terms of the mutual transconductance g_m. This describes the output signal current in terms of the input voltage signal:

$$g_m = \frac{i_C}{v_{BE}}$$

(see Figure 2.5.6).

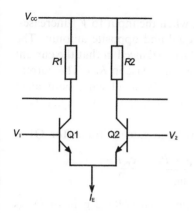

Figure 2.5.2 *The long tailed pair*

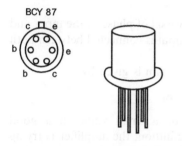

Figure 2.5.3 *A matched transistor pair*

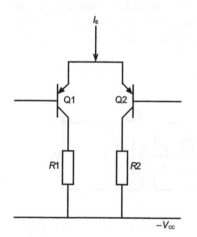

Figure 2.5.4 *A pnp BJT long tailed pair*

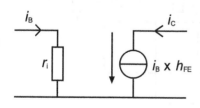

Figure 2.5.5 *The a.c. small signal model of a transistor*

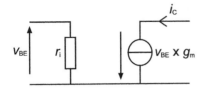

Figure 2.5.6 *Alternative transistor model with g_m*

In an ideal differential amplifier, when the input to $V1$ increases, the input to $V2$ decreases by an equal and opposite amount. The current in collector resistors is similarly balanced in that the current increase in R_{C1} is equal to the current decrease in R_{C2}. The current in the emitter resistor does not alter, so the small signal equivalent circuit of the long tailed pair with a differential mode signal is as shown in Figure 2.5.7.

In this circuit, the differential voltage gain of transistor Q1 is given by:

$$Av_{Q1} = \frac{v_o}{v_i} = \frac{g_m v_{BE} \times R_C}{\frac{1}{2} v_{BE}} = \frac{g_m R_C}{2}$$

and by a similar reasoning, the differential voltage gain of Q2 is given by:

$$Av_{Q2} = -\frac{g_m R_C}{2}$$

(The minus sign coming from the reversal of phase of the input and the $v_{BE}/2$ is due to the fact that the input is connected between both inputs.)

Hence the total differential voltage gain is given by:

$$Av = -g_m R_C$$

The common mode gain should be a small value in a good differential amplifier, because, by definition, the amplifier is trying to amplify the difference of the two input signals. For completeness, here is an analysis of the common mode amplifier gain.

In the previous circuit, changes in the collector currents balanced out, so the emitter resistor could be omitted. In this circuit, the voltage applied to the bases is the same, and so will affect the voltage in the emitter resistor. The equivalent circuit is slightly different, as was shown in Figure 2.5.7.

In the circuit of Figure 2.5.8, $v_{C1} = v_{C2} = -i_B \times h_{FE} \times R_C$.

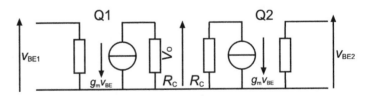

Figure 2.5.7 *Small signal equivalent circuit of long tailed pair with differential inputs*

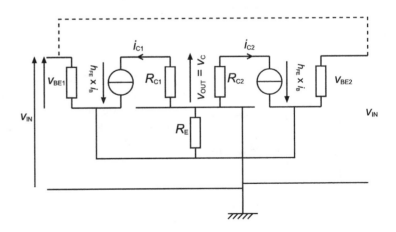

Figure 2.5.8 *A common mode amplifier equivalent circuit*

Summing the voltage drops from the input to ground gives:
$$v_{IN} = v_{BE} + (2 \times i_B + 2 \times i_B h_{FE}) R_E$$
This simplifies to:
$$v_{IN} = v_{BE} + 2 i_B (h_{FE} + 1) R_E$$
In terms of the output:
$$v_C = i_B h_{FE} R_C \quad \text{from which} \quad i_B = \frac{v_C}{h_{FE} R_C}$$
Combining these equations results in:
$$v_{IN} = v_{BE} + \frac{2 v_C}{h_{FE} R_C}(1 + h_{FE}) R_E$$
Now for the simplification:

(a) $h_{FE} \gg 1$
$$v_{IN} = v_{BE} + \frac{v_C (2 h_{FE}) R_E}{h_{FE} R_C} = v_{BE} + \frac{2 v_C R_E}{R_C}$$

(b) on the right-hand side of the equation, v_{BE} is very much smaller than the other term, so
$$v_{IN} = \frac{2 v_C R_E}{R_C}$$
$$\text{Gain} = \frac{v_C}{v_{IN}} = \frac{R_C}{2 R_E}$$

This final relationship tells us that the higher the emitter resistance, the smaller the common mode gain. This is why the better differential amplifiers (and all those used in operational amplifiers) have a differential stage which has a constant current source in the emitter circuit. These inherently have a high impedance.

The base–emitter comparator

On occasion, the circuit of Figure 2.5.9 can be used to amplify the difference in two a.c. signals. It does have its drawbacks in that the amplifier stage gain is somewhat limited, but it is still a viable circuit.

The output current is determined by the base–emitter voltage. In many amplifiers, the emitter is held at a constant voltage and the input signal applied directly to the base. In this circuit, the base and

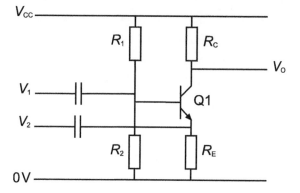

Figure 2.5.9 *The base–emitter voltage comparator*

the emitter of the transistor both have a.c. signals superimposed on the d.c. bias. The a.c. output will follow the mixer equation where:

$$v_{BE} = v_B - v_E$$

The operational amplifier mixer

The operational amplifier is the simplest circuit technique for comparing voltages (Figure 2.5.10):

$$V_o = A(V_2 - V_1)$$

This makes it look a very simple voltage comparison technique; but as you will see in Chapter 5, the input stage of an op amp is a version of a long tailed pair. The advantage is that the circuit design is very much more simple and involves far fewer components.

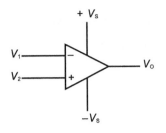

Figure 2.5.10 *The operational amplifier*

3 Power amplifiers

Summary

To some, a Power Amplifier is any device which takes a low power, low voltage audio signal, amplifies it, and outputs the result into a loudspeaker. This is, perhaps the most common application of the Power Amplifier, but it is by no means the only one. For example, many transducers need special power amplifiers to drive them: such as

    a.c. motor drives   r.f. aerial transmitters
    d.c. motor drives   c.r.t. scan coils
    solenoids   print heads
    servo mechanisms

This chapter looks into the methods used by designers of electronic circuits to drive power into a load.

3.1 Types of amplifier

Amplifiers take an input signal and increase the value of the input voltage and/or the input current to provide a gain in voltage, current and (usually) output power. These gains have the symbols:

A_v voltage gain
A_i current gain
A_p power gain

Figure 3.1.1 shows a schematic of a general purpose amplifier.

Amplifiers are also characterized both by the amount by which they boost the signal and the frequency range over which they operate. These parameters are called the gain and the bandwidth.

Amplifier gain and frequency response

Figure 3.1.2 shows a band pass amplifier. This is a class of amplifier which amplifies a **band** of frequencies. In the response shown, the

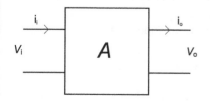

Figure 3.1.1 *Schematic of a general purpose amplifier*

Figure 3.1.2 *An amplifier frequency response*

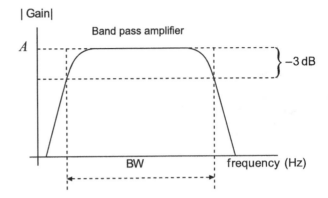

amplifier will not amplify very low or very high frequencies. It attenuates (or cuts off) these frequency components.

Characteristics such as these allow you to identify two main parameters which are common to all amplifiers:

(i) Mid-band gain A. This is the gain in the centre of the frequency response. It could be expressed in decibels or as a ratio of output to input.
(ii) Bandwidth. This is the range of frequencies over which the amplifier operates. It is normally measured between the $-3\,\text{dB}$ roll-off points as shown in the figures. $-3\,\text{dB}$ is the equivalent of measuring when the voltage gain has fallen to 0.707 of the mid-band value.

Mathematics in action

The Bel is the unit of power ratios which was found useful by telecommunication engineers. It was named after Alexander Graham Bell, and was introduced to account for the large dynamic range between input and output powers. A Bel is defined as:

$$1\ \text{Bel} = \log_{10} \frac{\text{Power out}}{\text{Power in}}$$

As the Bel unit was an inconvenient size, the decibel was found to be more convenient:

$$1\ \text{decibel} = 1\ \text{dB} = 10 \log_{10} \frac{P_o}{P_i}$$

This was originally used for dealing with power losses of transmission lines and then the power gains of amplifiers and systems. The unit was so useful that electronics engineers adopted it for electronic circuits.

$$\text{Power} = \frac{V^2}{R} \quad \text{or, more specifically,}$$

$$P_o = \frac{V_o^2}{R_o} \quad \text{and} \quad P_i = \frac{V_i^2}{R_i}, \quad \text{so:}$$

$$1\ \text{dB} = 10 \log_{10}\left(\frac{V_o^2/R_o}{V_i^2/R_i}\right)$$

Power amplifiers

In the balanced systems used in the world of telecommunications, $R_o = R_i$ so this simplified the relationship to:

$$1\,\text{dB} = 10\log_{10}\frac{V_o^2}{V_i^2} = 20\log_{10}\frac{V_o}{V_i}$$

It often occurs that through a mathematical laziness, electronic engineers conveniently forget the dependence on input and output resistance matching and use the voltage version of the decibel formula in its simplest form.

Example 3.1.1

In Figure 3.1.3(a), evaluate:

(i) mid-band gain;
(ii) bandwidth.

The mid-band gain is read off the scale to be 50:

$$20\log_{10} 50 = 34\,\text{dB}$$

The roll-off points are measured at $-3\,\text{dB}$ from the mid-band gain:

$$34 - 3 = 31\,\text{dB}\;\text{which is a gain of approximately 35}$$

From the graph, the $-3\,\text{dB}$ points occur at 30 kHz and 100 kHz. The bandwidth is thus $100 - 30 = 70\,\text{kHz}$.

Problem 3.1.1

For each of the frequency response curves in Figure 3.1.3(b), (c), (d) evaluate:

(a) the mid-band gain (in dB and as a gain term);
(b) the bandwidth (in Hz).

Logarithmic axes

For simplicity, all of the problems in Figure 3.1.3 were drawn to linear axes. In practice, nearly all frequency responses are drawn to a logarithmic scale, as shown in Figure 3.1.4. The feature of logarithmic axes is that decades of frequency such as 10 Hz, 100 Hz, 1 kHz, 10 kHz, etc. are equal distances apart on the graph.

The tuned amplifier

The tuned amplifier is not really a special case of the band pass amplifier. The tuned centre frequency is usually the result of

Figure 3.1.3 Amplifier responses

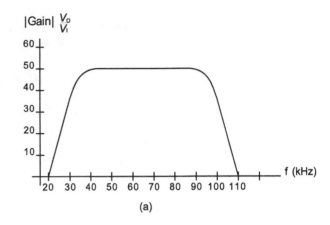

(a)

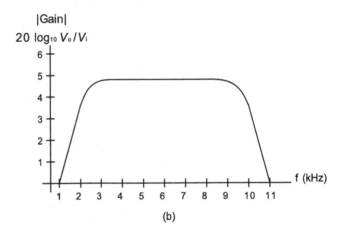

(b)

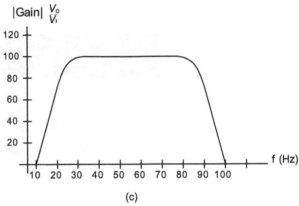

(c)

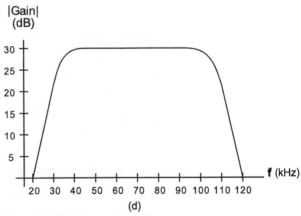

(d)

Figure 3.1.4 *A logarithmic frequency axis*

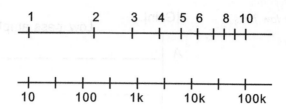

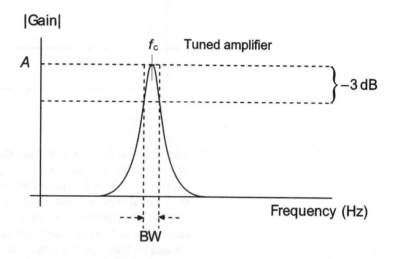

Figure 3.1.5 *A tuned amplifier*

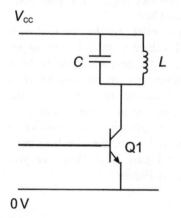

Figure 3.1.6 *A resonant LC output stage of an amplifier*

resonance between inductive and capacitive components (Figure 3.1.5). Figure 3.1.6 shows a typical output stage which would produce this response. Altering L or C shifts the centre frequency and usually alters upper and lower cut-off frequencies together. To a reasonable approximation, a parallel LC circuit will resonate at a frequency of:

$$f_o = \frac{1}{2\pi\sqrt{LC}}$$

A practical inductor contains resistance which absorbs resonant energy and so changing this will broaden or narrow the bandwidth.

Band pass, low pass and high pass amplifiers

A band pass amplifier has separate components which independently adjust the lower and upper cut-off frequencies. It can only amplify a range of frequencies. The application determines the ranges of these frequencies. You have already worked through a few different types of band pass response in Problem 3.1.1. Different frequency bands have different band names. Here are some in common use:

Audio (AF)	20 Hz	20 kHz
Low frequency (LF)	30 kHz	300 kHz
Medium frequency (MF)	300 kHz	3 MHz
Video amplifiers	d.c.	10 MHz
High frequency (HF)	3 MHz	30 MHz
Very high frequency (VHF)	30 MHz	300 MHz
Ultra high frequency (UHF)	300 MHz	3 GHz

Figure 3.1.7 *The low pass amplifier*

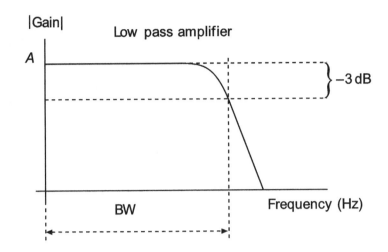

The low pass amplifier has the capability of amplifying d.c. voltages. For example, a 2 mV signal input into an amplifier with a gain of −100 will produce an output of −0.2 V. This is necessary in motor drives or heater outputs, but not always so vital for audio signals. The main disadvantage of d.c. amplifiers is the degradation caused by drift as the amplifier heats up or even as its components change as they age. Care must be taken to ensure that the amplifier does not introduce unwanted d.c. terms (see page 35). Figure 3.1.7 shows the basic shape of the low pass filter.

High pass amplifiers do not really exist. In the limit, if you measure higher and higher frequencies there is always some upper cut-off point. A high pass amplifier could be considered to be a special case of a band pass amplifier where you are not so concerned about the upper cut-off frequency. Of course, there is always the problem of noise, and some of the more exotic forms of amplifier distortion. These can be made worse by allowing the amplifier to operate at excessively high frequencies. In practice, the frequency response is similar to that of those in Figure 3.1.3. However, you will usually see them drawn like that in Figure 3.1.8.

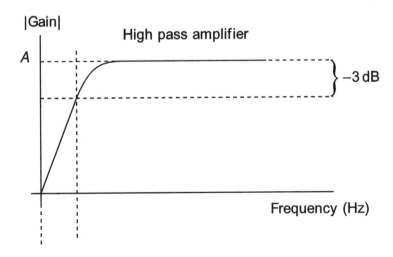

Figure 3.1.8

Power amplifiers

Figure 3.1.9 *A clipped sine wave*

An audio problem, or the tweeter-buster

If an amplifier is overdriven, the output signal tries to exceed the supply voltage and the waveform is clipped (Figure 3.1.9).

A sine wave contains only one frequency ... called its **fundamental**. When the sine wave is clipped, a frequency analysis would show that it contains substantial energy not only at the fundamental frequency (called f_0), but also at the odd harmonics $3f_0$, $5f_0$, $7f_0$, etc. A 1 kHz sine wave when clipped would generate waveforms at 3 kHz, 5 kHz, 7 kHz, etc. Tweeters are the high frequency output units of loudspeakers. So, if a waveform is clipped, these high frequency harmonics can overdrive the tweeter and damage it. It is a well understood failure mechanism of tweeters. Moral: don't overdrive your amplifier into distortion!

3.2 Classes

Amplifiers are divided into **classes**. Each class describes how the output power devices handle the current delivered to the load. These devices could be bipolar transistors (BJTs), field effect transistors (FETs) or valves (yes, they are still popular in some applications). They can be biased to operate in what is termed Class A, B or C. These are summarized in the next three paragraphs before being analysed more thoroughly in the rest of Section 3.2.

Class A

The circuit of Figure 3.2.1 is operating in Class A. The active device Q1 conducts all of the time. Ideally, with no input signal present, Q1 would be biased at a value to maintain point X to 10 V (which is half the supply voltage). Thereafter, when an input signal was applied, the output could swing to the maximum extent as shown in Figure 3.2.2. The efficiency of these circuits is very low since Q1 conducts even when no signal is present. Current flows though R1 in order to produce a voltage drop of 10 V so, even with no signal, R1 and Q1 will dissipate energy in the form of heat.

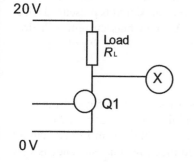

Figure 3.2.1 *Class A output stage*

Class B

Figure 3.2.3 requires two output devices as a minimum to work in Class B. With no input current, no current will flow through the

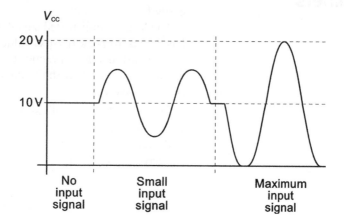

Figure 3.2.2 *The voltage at point X for (i) no bias; (ii) small input signal; (iii) maximum input signal*

Figure 3.2.3 *A Class B output stage schematic*

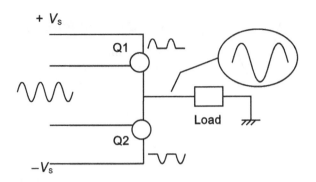

path from $+V_{CC}$ through the active devices to the $-V_{CC}$ supply. When an a.c. signal is applied, each active device conducts for 50% of the time: Q1 conducts for the positive half cycles and Q2 for the negative half cycles. Confusingly, it is possible to use this configuration for Class A (as we shall do shortly). Remember – in Class A, both output devices must conduct regardless of the output voltage. In Class B, each output device only handles 50% of the signal.

There is a technical problem with this method of supplying output power which we shall investigate shortly. The waveform shown is not strictly correct since all active devices require a small voltage to make them start conducting. For this reason, Class AB was developed. In this class, when no input signal is present, devices Q1 and Q2 both conduct slightly. A small bias current will flow through them from the $+V_{CC}$ supply to the $-V_{CC}$ supply.

Class C

In Class C, the active device conducts for less than 50% of the time. It is used almost always for r.f. amplifiers which have an LC 'tank' circuit as a load. The oscillations are maintained in the LC circuit even when the active device is not conducting. The efficiency is very high.

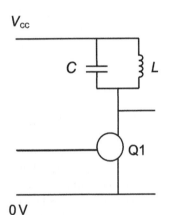

Figure 3.2.4 *A Class C output stage*

3.3 Class A amplifiers

Most small signal amplifiers operate in class A. For the configuration which was shown in Figure 3.2.1, the efficiency is low and can easily be worked out.

The largest undistorted signal which can be delivered to the load has a peak-to-peak (pk-pk) value approaching $V_{CC} - 0$ volts. (This is rarely attainable in practice because of Q1's inability to maintain a linear output when the voltage across it falls near zero, i.e. the signal is distorted.)

The largest peak voltage is thus $V_{pk} = V_{CC}/2$ and the largest RMS voltage is $V_{RMS} = (V_{CC}/2\sqrt{2})$.

Power is simply $P = (V^2/R)$, so the maximum a.c. power delivered to the load is:

$$P = \left(\frac{V_{CC}}{2\sqrt{2}}\right)^2 \frac{1}{R_L} = \frac{1}{8}\frac{V_{CC}^2}{R_L}$$

The average power delivered to the circuit is simply $P\,dc = I_C \times V_{CC}$ where I_C is given by $\frac{\frac{1}{2}V_{CC}}{R_L}$. This makes $Pdc = \frac{1}{2}\frac{V_{CC}^2}{R_L}$.

So, the highest efficiency of the Class A amplifier shown is:

$$\eta = \frac{\frac{1}{8}V_{CC}^2 R_L}{\frac{1}{2}V_{CC}^2 R_L} = 25\%$$

Problems 3.3.1

(1) The waveform shown was measured at point Y in Figure 3.3.1:
   (a) What power is delivered to the load?
   (b) At what efficiency is the amplifier working?
(2) If the output waveform in Figure 3.3.1 falls to 4 V peak-to-peak, recalculate:
   (a) the power delivered to the load;
   (b) the amplifier efficiency.
(3) Use the formulae you have been using and input them into either:
   (a) a MathCAD type program; or
   (b) a spreadsheet.

Use these to construct a table and then a graph to show the relationship between:

   (i) output pk-pk voltage and power output;
   (ii) output pk-pk voltage and efficiency.

D.c. blocking

Most audio amplifiers operate into inductive loudspeakers, so the arrangement of Figure 3.2.1 cannot be used.

Figure 3.3.2 shows a schematic of the internal construction of a loudspeaker. As the a.c. signal passes through the voice coil, the electromagnet thus created reacts with the permanent magnet. This

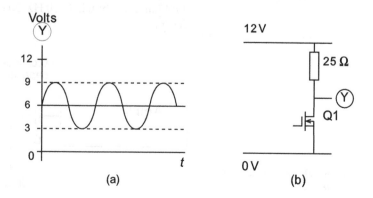

Figure 3.3.1 *A Class A output stage*

66 Higher Electronics

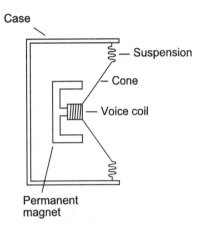

Figure 3.3.2 *A loudspeaker*

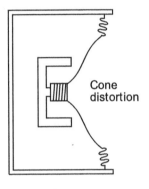

Figure 3.3.3

drives the cone backwards and forwards and converts an electrical signal into an acoustic signal. Problems occur if a d.c. signal is allowed to pass through the voice coil. It still becomes an electromagnet, but the d.c. causes an offset in the voice coil position. There are now three possible errors:

(i) The d.c. current will cause a heating effect, which when an a.c. signal is superimposed on the d.c. could cause the voice coil to overheat and fail.
(ii) There is now additional tension in the cone. It will mechanically deform and produce a distorted signal.
(iii) An a.c. signal superimposed on the d.c. could cause the voice coil to try to move beyond the limit of the movement, e.g. it could strike the magnet assembly, causing gross distortion and/or damage to the loudspeaker.

Of these, error (ii) is the most likely to occur. Figure 3.3.3 shows the possible effect of the distortion.

Since a d.c. bias through the loudspeaker voice coil is held not to be a good thing, then it must be prevented. This can be achieved by:

(a) electronic monitoring and control;
(b) a.c. coupling techniques with:
   (i) a transformer;
   (ii) a capacitor.

The simplest and cheapest method uses a capacitor as in Figure 3.3.4.

Capacitive coupling

What about a value for *C*? Most domestic loudspeakers have a nominal impedance of $8\,\Omega$. *C* and LS of Figure 3.3.4 form a high pass filter, so if we want to calculate a suitable value of *C*, then we must investigate the expected frequency response of the system. Earlier, using Figure 3.1.2, an amplifier was shown to have a typical frequency response as shown again in Figure 3.3.5.

Common wisdom sets a reasonable quality hi-fi frequency response lying between $50\,\text{Hz}$ (f_L) and $20\,\text{kHz}$ (f_H). This is, of course, very subjective and the letters pages of many an Electronics or Audio magazine often will set a debate going about the ranges of perception of the human auditory system. For many not in the 'Golden Ear' bracket, $50\,\text{Hz}–20\,\text{kHz}$ is a fair guideline for audio design.

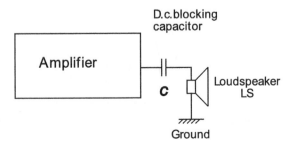

Figure 3.3.4 *An a.c. coupled load using a capacitor*

Power amplifiers 67

Figure 3.3.5 *A typical audio amplifier response*

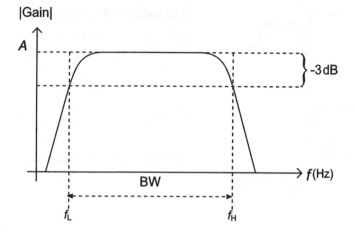

Example 3.3.1

Find the value of capacitor which is suitable for passing frequencies down to 50 Hz into an 8 Ω loudspeaker.

Our capacitor must block d.c. but allow frequencies above 50 Hz through. That is, 50 Hz represents the −3 dB roll-off frequency where its reactance $|X_C|$ equals the load impedance.

Recall that $X_C = \dfrac{1}{2\pi f C} = 8\,\Omega$ at 50 Hz so that:

$$\dfrac{1}{2\pi 50 C} = 8\,\Omega$$

$$C = \dfrac{1}{2 \times \pi \times 50 \times 8} = 398\,\mu F$$

Many electrolytic capacitors are only manufactured to a tolerance of ±20%, so a 470 μF is probably called for.

Problems 3.3.2

(1) Calculate the value of the d.c. blocking capacitor which would be suitable for a lower frequency response down to 20 Hz in to an 8 Ω speaker.
(2) Find the d.c. blocking capacitor needed to allow a 4 Ω loudspeaker to operate down to 30 Hz.

The Class A circuit needed to operate with the d.c. blocking capacitor would be amended most simply to that of Figure 3.3.6.

At a frequency where X_C is negligible, the a.c. load impedance would now be the parallel combination of R_C and R_L. The collector resistor is not now the load, but just another circuit element capable of dissipating power. Hence with this circuit, the efficiency falls to a maximum of 12.5%.

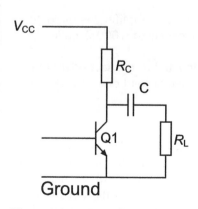

Figure 3.3.6 *Class A amplifier with a.c. coupled output*

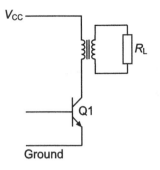

Figure 3.3.7 *Class A amplifier with transformer coupled load*

Transformer coupling

The other way to block d.c. is to use a coupling transformer as in Figure 3.3.7. R_L is not now part of the collector load, so there is no d.c. power loss through it and the efficiency of this circuit rises to 50%.

Another useful feature is the impedance matching capability of transformers, so that by choosing a suitable turns ratio, a collector load in the kΩ range can be perfectly matched to a load (of say 16 Ω).

Transformer matching

$$\frac{v_P}{v_S} = \frac{N}{1} \qquad v_P = N \times v_S$$

$$\frac{i_P}{i_S} = \frac{1}{N} \qquad i_P = \frac{1}{N} \times i_S$$

Dividing gives:

$$\frac{v_P}{i_P} = \frac{N}{1/N} \times \frac{v_S}{i_S}$$

$$\frac{v_P}{i_P} = N^2 \times \frac{v_S}{i_S}$$

or $Z_P = N^2 Z_S$.

Example 3.3.2

What turns ratio of coupling transformer would be used to match a 16 Ω load to an expected 25 kΩ collector load?

$$Z_P = N^2 \times Z_S$$

$$1 \text{ k}\Omega = N^2 \times 16$$

$$\frac{2500}{16} = N^2$$

$$N = 12.5$$

Problems 3.3.3

(1) Calculate the turns ratio of an amplifier required to match an 8 Ω load to an expected 1000 Ω collector load.
(2) If a 4 Ω loudspeaker is connected to the secondary of a 20:1 transformer, what collector load does it present?

Transformers can be bulky and expensive. The core material is determined by the frequency at which the transformer is to operate. Audio requires some very low frequencies, and to make the magnetic circuit efficient and low loss increases the weight of the transformer. Nevertheless, they are still used in some valve-driven audio amplifiers. At radio frequencies for transmitter circuits, the core materials are much lighter, so it becomes more cost effective to use transformers.

Electronic detection of d.c. bias current in the voice coil depends on an integrating amplifier. If, over a period of time (of the order of seconds), it detects an overall current through the load, it adjusts the bias of some of the amplifier components to reduce the d.c. offset to zero. Directly coupled audio amplifiers have the advantage of operating down to low audio frequencies – a feature which enhances the fidelity of reproduction. (Bear in mind, though, that even 'good' bass subwoofers have a frequency response which only goes down to around 10–15 Hz.)

Class A push–pull design

In a push–pull amplifier output stage, there are two active devices as shown in Figure 3.3.8(a) and (b).

The active devices could be valves, MOSFETs or bipolar junction transistors; they could be the same type of device or complementary (i.e. 1 npn or 1 pnp). Design (a) shows a dual rail amplifier, while design (b) is a single rail amplifier. In both cases, we assume that we are dealing with an audio amplifier driving a loudspeaker.

The ideal voltage at point X in Figure 3.3.9 would be 15 V and C would undoubtedly be an electrolytic capacitor mounted with the polarity as shown. In normal use, Q1 and Q2 would both be conducting, maintaining $V_{CE(Q1)} = 15$ V and $V_{CE(Q2)} = 15$ V.

This means that the d.c. bias voltages are:

$$V_{B1} = 16.2\,\text{V} \quad V_E = 15\,\text{V} \quad V_{BE} = 1.2\,\text{V}$$
$$V_{B2} = 13.8\,\text{V} \quad V_E = 15\,\text{V} \quad V_{BE} = 1.2\,\text{V}$$

When the signal goes positive, V_{B1} and V_{B2} waveforms alter as shown. To remain as a Class A amplifier, both Q1 and Q2 must conduct continually, i.e. never being cut off. $V_{CE(Q1)}$ reduces while $V_{CE(Q2)}$ increases for a positive change of output voltage, and $V_{CE(Q1)}$ increases while $V_{CE(Q2)}$ reduces for a negative change of output voltage. The d.c. voltage supply to a Class A amplifier

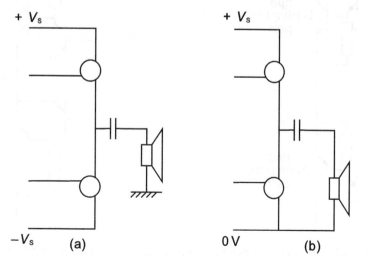

Figure 3.3.8 *Class A push–pull output stages*

Figure 3.3.9 *A Class A output stage*

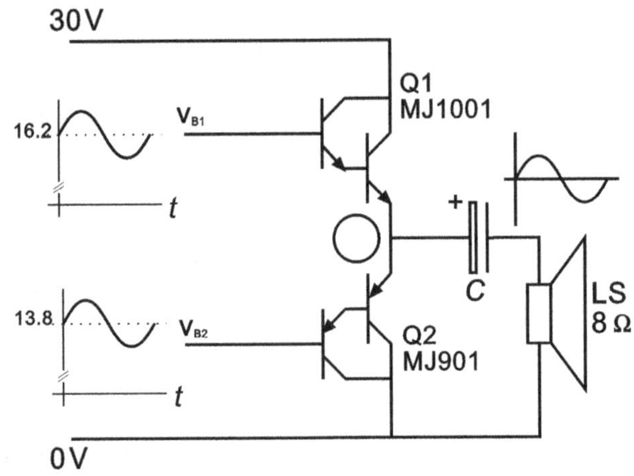

remains essentially constant, irrespective of the output power being delivered to the load.

Darlington pairs

The npn and pnp Darlington transistor configurations are those most commonly used. Current gains of 1000 are easily achieved with devices which can handle up to 50 A. Occasionally, complementary Darlington transistors are used (Figures 3.3.11–3.3.13).

In conventional Darlington pairs, the resistor R prevents the leakage current which flows through Q1 from biasing Q2 on.

The output stage of Figure 3.3.9 has two inputs which need to be driven by some transistor stage. Figure 3.3.14 shows a more complete output stage. In this circuit, Q1 and Q2 need the same drive signal since (during the first half cycle shown) Q1 will turn ON more and Q2 will turn OFF more.

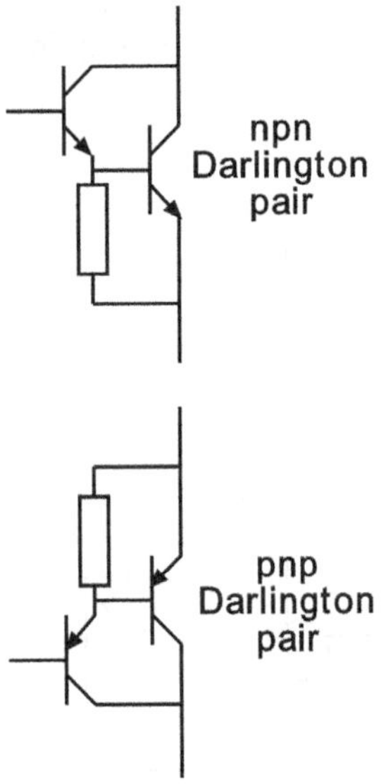

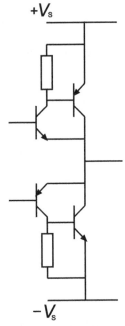

Figure 3.3.10 *npn and pnp Darlington pairs*

Figure 3.3.11 *Complementary Darlington pairs*

Power amplifiers

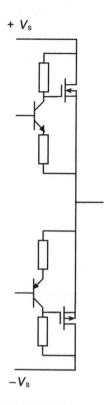

Figure 3.3.12 *Mixed BJT/MOSFET combinations ready to create a complementary output*

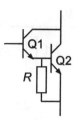

Figure 3.3.13 *Darlington pair with leakage resistor*

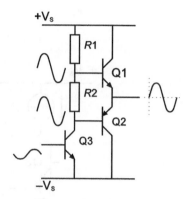

Figure 3.3.14 *Class A output with driver transistor*

Another configuration uses the same type of transistor (npn in the case of Figure 3.3.15). Q1 is a non-inverting emitter follower while Q2 is an inverting common emitter amplifier. Since these devices are now of the same polarity, a different tactic for driving them is required. Transistor Q3 in Figure 3.3.16 is called a **phase-splitter**. If R1 = R2, then collector and emitter voltages will be equal and opposite. When Q1 is driven ON more; then Q2 will be driven OFF more. This would correspond to the output at its first positive half cycle as shown in the waveforms. This is different from the drive circuit needed for complementary output transistors in Figure 3.3.14.

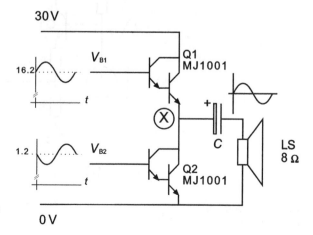

Figure 3.3.15 *An alternative Class A output stage*

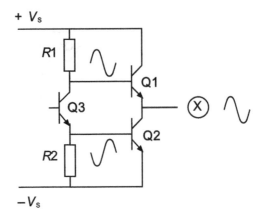

Figure 3.3.16 *Class A output with phase splitter*

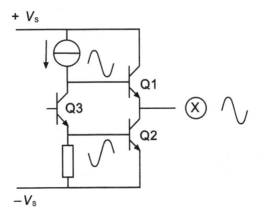

Figure 3.3.17 *A constant current source as the collector load*

In both of these circuits, references to 'turning ON more' and 'turning OFF more' imply a small change of base–emitter voltage bias which alters the value of V_{CE} to change the output voltage.

The performance of both of these circuits can be improved with either or both of the circuits shown in Figures 3.3.17 and 3.3.18.

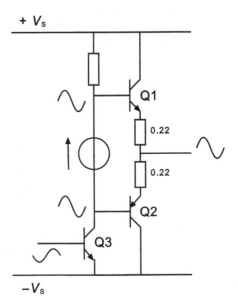

Figure 3.3.18 *A constant voltage (bias) source between the bases*

Figure 3.3.19 *Amplifier circuit diagram*

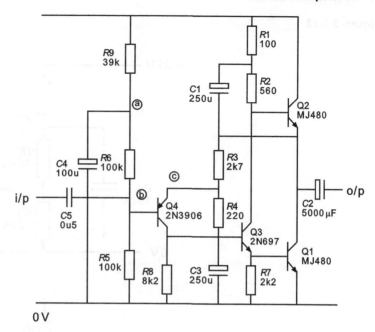

Case study – the Linsley-Hood Class A amplifier

This classic of the specialist audio amplifier designer John Linsley-Hood first appeared in 1969. The output stage is a variant on Figure 3.3.16, where Q3 acts as a phase splitter. *C*1 is a large electrolytic capacitor which is charged to a d.c. voltage by *R*1. It maintains a constant voltage across *R*2. When a constant voltage acts across a resistor, it becomes the constant current source of Figure 3.3.17. Remember, it is not possible to instantaneously change the voltage across a capacitor – point X is the same as Y which maintains a constant voltage across *R*2.

A.c. feedback is provided through *R*3 and *R*4, as shown in Figure 3.3.21. The feedback ratio is:

$$\beta = \frac{R4}{R4 + R3} = \frac{220}{220 + 2700}$$

The a.c. gain $\approx 1/\beta \approx 13$. Q4 acts as the comparison element.
The published performance is:

Power output	10 W
Distortion	< 0.05%
Frequency response	30 Hz–180 kHz

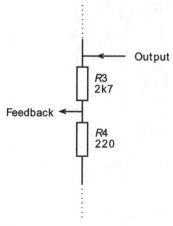

Figure 3.3.20 *The bootstrap capacitor C1*

Figure 3.3.21 *The feedback components*

Problems 3.3.4

(1) Calculate the base voltage at b in Figure 3.3.22.
(2) Since V_{BE} is typically 0.6 V, calculate the currents c and d in Figure 3.3.23. (At d.c. no current will flow through *R*4, so assuming the base current of Q3 is negligibly small, all current d will flow through *R*3.)
(3) Calculate V_{R3} in Figure 3.3.24.

74 Higher Electronics

Figure 3.3.22

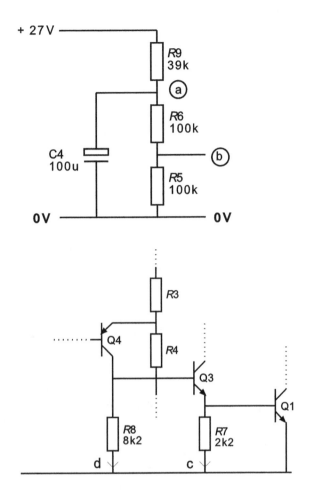

Figure 3.3.23

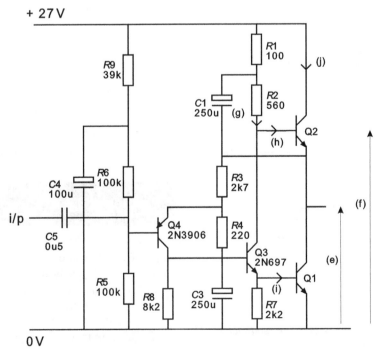

Figure 3.3.24

(4) Hence calculate the output voltage e and Q1 base voltage f in Figure 3.3.24.
(5) Calculate collector current g in Figure 3.3.24. Some of this passes through R7, but the rest is equally divided between the bases of Q1 and Q2 (if they are reasonably well matched).
(6) Calculate the currents h and i in Figure 3.3.24.
(7) MJ480s might have an h_{FE} typically of 120. Calculate the standing current in the output circuit and hence the quiescent power dissipated by the output stage.

3.4 Class B amplifiers

Figure 3.4.1 shows how to implement a Class B output with bipolar junction transistors, MOSFETS and valves. In true Class B, when no signal is present, neither Q1 nor Q2 conducts. When the signal becomes positive, Q1 conducts to supply current to the load. When the input becomes negative, Q2 conducts to supply current to the load. Unfortunately, no real active devices exist with a characteristic which starts conducting as soon as the input rises above 0 V. A BJT requires around 0.6 V at V_{BE} before it will conduct, while a MOSFET could require up to 4 V V_{GS}. Put another way, all practical devices will not start conducting until the input reaches a particular threshold.

In each of these circuits, Q1 conducts during the positive half cycles and Q2 during the negative half cycles.

Crossover distortion

In Figure 3.4.2, the bases of the transistors are driven simply from a common input. However, as BJTs, Q1 and Q2 need approximately 0.6 V before they start conducting. The output waveform (Figure 3.4.3) shows the typical **crossover distortion** which results from the V_{BE} threshold of BJTs.

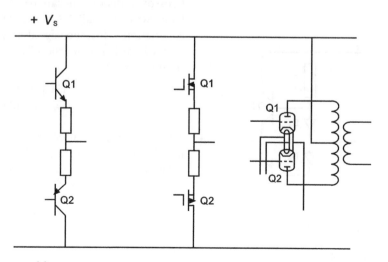

Figure 3.4.1 *Schematic of a Class B amplifier*

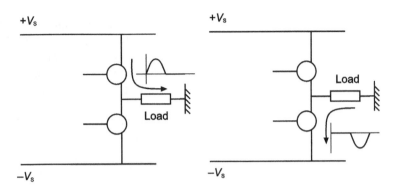

Figure 3.4.2 *Current flow in Class B*

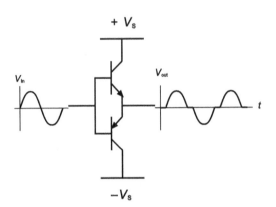

Figure 3.4.3 *Crossover distortion*

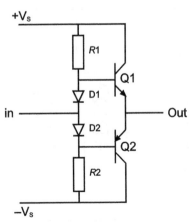

Figure 3.4.4 *Biasing into Class AB*

Class B output bias

The threshold of MOSFETs is somewhat less predictable between devices, so we shall confine our initial development to BJTs.

The $V_{BE} = 0.6$ arises from the base–emitter diode voltage drop. Crossover distortion can be eliminated by biasing the transistor bases with a small 0.6 V source. Since they are then forced into conduction when no signal is present, the class of operation is somewhere between true Class B and Class A – called Class AB (Figure 3.4.4):

D1 provides bias for Q1
D2 provides bias for Q2

The disadvantage, of course, is when the diode V_F does not exactly match the transistor V_{BE}. The commonest solution is to replace D1 and D2 with the configuration called a 'V_{BE} multiplier' or more simply 'a rubber Zener'. This is the circuit of Figure 3.4.5(a) and (b).

Setting up is normally achieved with a potentiometer as shown in (a), but for ease of calculation, use (b):

$$V_{BE} = 0.6\,\text{V} \quad \text{and} \quad I_{R2} = \frac{V_{BE}}{R_2}$$

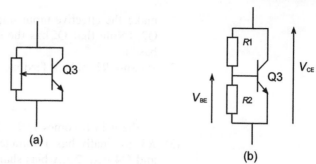

Figure 3.4.5 Bias circuits for Class AB

Assuming that I_B is negligible:

$$I_{R2} = I_{R1}$$

so $V_{CE} = I_{R2} \times (R_1 + R_2)$

$$V_{CE} = V_{BE} \times \frac{R_1 + R_2}{R_2}$$

The idea is that V_{CE} can be adjusted by the potentiometer, its stability being set by the stability of V_{BE}.

Class B case study

Here is a 1982 FET design by Peter Wilson of International Rectifier plc. The published performance figures for this amplifier are:

Distortion	< 0.01%
Bandwidth	15 Hz to 100 kHz (-1 dB points)
Gain	22
Output power	32 W

Working from input to output in the circuit (i.e. left to right):

(1) Q1 and Q2 form a differential (long tailed) pair which amplifies the difference between the input and a proportion of the output. R1 and R2 make a small input circuit which helps to

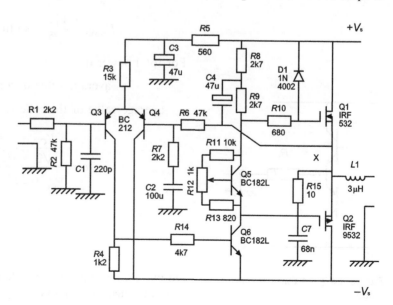

Figure 3.4.6 A practical FET-based Class B amplifier

make the effective input impedance to Q1 the same as that of Q2. (Note that Q2 has the same components connected to its base.)

(2) R6 and R7 make a feedback factor of:

$$\beta = \frac{2.2}{2.2 + 47} = 0.045$$

so the gain becomes $1/\beta = 22$.

(3) R3 nominally has a voltage of 30 V across it, so provides Q3 and Q4 with 2 mA bias shared between them.

The voltage drop across R4 becomes $1\,\text{mA} \times 1200 = 1.2\,\text{V}$. This provides adequate current into the base of Q6 to set the quiescent point of the circuit. The bias of Q1 and Q2 will be set so that point X is at approximately 15 V. The precise value is that which sets the base voltages of Q3 and Q4 at the same level.

(4) Q5, R11, R12 and R13 provide the bias adjustment for Q1 and Q2.

(5) C4 provides constant current feed to the bias circuit by maintaining the voltage across R9 constant irrespective of the voltage drive on the gate of Q1.

(6) C1, R15 and C7 limit the high frequency response. Further trimming is often achieved by connecting a small (47 pF?) capacitor between the base and collector of Q6.

Efficiency of Class B

Figure 3.4.7 shows a simplified schematic of the output stage. The efficiency η is given by:

$$\eta = \frac{\text{useful (a.c.) power out}}{\text{total power in}}$$

For a.c. power out:

$$\text{maximum voltage across } R_L = V_{CC}$$
$$\text{maximum signal current through } R_L = V_{CC}/R_L$$

$$V_{RMS} = \frac{V_{CC}}{\sqrt{2}} \quad \text{so that} \quad I_{RMS} = \frac{V_{CC}}{\sqrt{2} \times R_L}$$

For total power in:

$$\text{average value of a sine wave is } V_{PK} \times 0.636$$
$$\text{average value of the collector current is } (V_{CC} \times 0.636)/R_L$$

$$\eta = \frac{\frac{V_{CC}}{\sqrt{2}} \times \frac{V_{CC}}{\sqrt{2} \times R_L}}{V_{CC} \times \frac{V_{CC} \times 0.636}{R_L}} \times 100\%$$

$$= \frac{1}{2 \times 0.636} \times 100\%$$

$$= 78.5\%$$

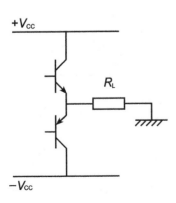

Figure 3.4.7 *A Class B output stage*

Output swings up to the voltage rails are neither normal nor recommended. When operating at non-maximal amplitudes, the

Power amplifiers

efficiency will be lower than 78.5%. When biased into Class AB, the efficiency will fall, of course, since the amplifier output devices are once more passing bias current even when no signal is present.

Problems 3.4.1

(1) Why is it not recommended to allow the output devices to operate near, or at, the supply rails?
(2) If $+V_{CC} = 20\,V$ and $-V_{CC} = -20\,V$ calculate the efficiency of an amplifier when the output amplitude is 20 V pk-pk (Figure 3.4.8).
(3) Calculate the efficiency of a Class B amplifier operating from supplies of $\pm 25\,V$ when the output is $13\,V_{RMS}$.
(4) A Class B amplifier has a gain of 50 and is powered from bipolar 40 V supplies (i.e. $+40\,V$ and $-40\,V$). Calculate the efficiency when the input is $0.35\,V_{RMS}$.
(5) Figure 3.4.9 shows the circuit of a Class AB amplifier:
 (a) What value of current is produced by the constant current source of R_3, Q_9, D_1, D_2, R_6?
 (b) What is the closed loop gain of the circuit?
 (c) What is the function of C_3; what would be the effect of doubling it?
 (d) (i) What is the quiescent value of voltage on the base of Q_3?
 (ii) Hence calculate the bias current flowing through R_{12}/R_{13}.
 (e) What bias voltage should be set at V_{CE} for Q5?

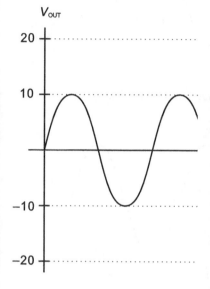

Figure 3.4.8

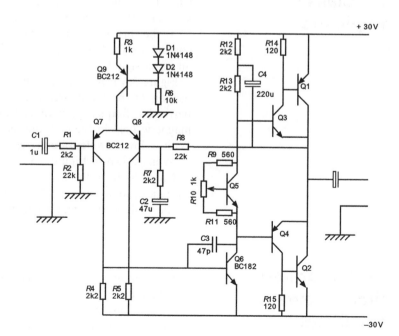

Figure 3.4.9

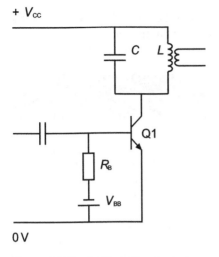

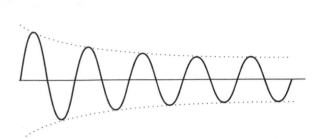

Figure 3.5.1 *A Class C output stage*

Figure 3.5.2 *Class C output collapsing*

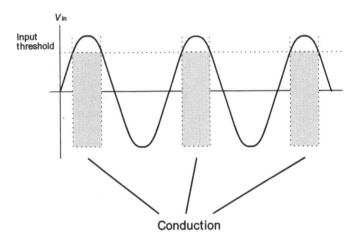

Figure 3.5.3 *The conduction angle of a Class C amplifier transistor*

3.5 Class C amplifiers

Class C amplifiers are very efficient. The active devices conduct for less than 50% of the input signal. The implication is that the device must be biased so that the input signal only forces it into conduction for less than 50% of the time for which it is amplifying an input signal.

Figure 3.5.1 shows a very simple Class C gain stage. Note that it uses an *LC* resonant circuit. This 'tank' circuit will resonate at one particular frequency, with energy alternately being stored as a magnetic field in the coil and as an electric field in the capacitor. With a low parallel circuit resistance, the oscillations would die away as shown in Figure 3.5.2.

The drive from the transistor is the equivalent of giving a child's swing a push at the correct instant. The oscillations are maintained in the tank with minimal energy expended from the drive circuit (Figure 3.5.3).

3.6 Integrated circuit power amplifiers

Power amplifiers consist of many transistors, and it was historically not long before semiconductor manufacturers started integrating them onto one piece of silicon. To start with, they had a dubious audio performance, but devices now are available from 0.5 W to

150 W; all able to operate at no worse than 0.1% total harmonic distortion. This is not the *Hi*-est of *Fi*, but it is certainly adequate for many domestic and commercial devices which need an audio output. IC amplifiers are starting to challenge discrete designs in terms of performance. However, the very best amplifiers still use discrete components.

Monolithic amplifiers have a wide range of applications in telephones, intercoms, televisions, radios, audio cassette recorders and many other devices. In this section we will look at a cross-section of **four** commercial devices, as shown in Table 3.6.1.

Table 3.6.1

Device	TBA820M	TDA2030V	TDA2050V	LM12CLK
Output power	1.6 W	8 W	25 W	150 W
Supply voltage	3–16 V	± 15 V	38–45 V	± 30 V
Bandwidth	20 kHz	140 kHz	20 kHz	60 kHz
Distortion	0.4% @ 0.5 W	0.1% @ 4 W	0.05% @ 15 W	0.01% @ 150 W

Source: Maplin MPS

The circuits in Figures 3.6.1–3.6.4 show the application circuits for each of these.

As always with integrated circuits for specific purposes, there is not much flexibility in the application circuits which use them. The manufacturers normally make them easy to use with minimum pin connections. Only the Pentawatt package used by examples TDA2030 and TDA2050 have a common output pin function.

Bridges

One way to increase output power is to use power amplifiers in a bridge configuration (Figure 3.6.5). Instead of connecting one end of the loudspeaker to ground, it is connected to two amplifiers acting in antiphase. This doubles the voltage across the L/S. The

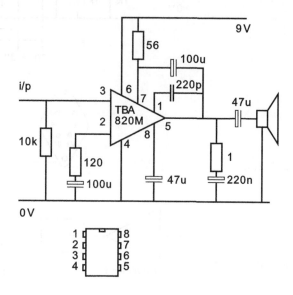

Figure 3.6.1 *The TBA820M*

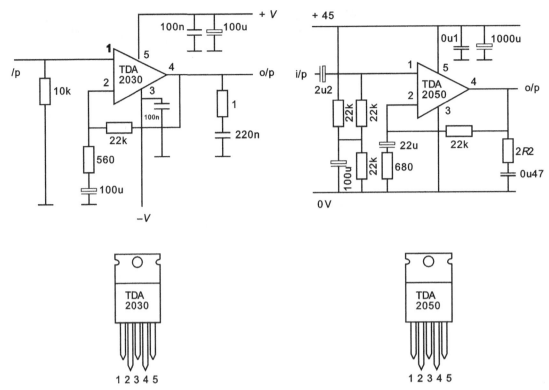

Figure 3.6.2 *The TDA2030V*

Figure 3.6.3 *The TDA2050V*

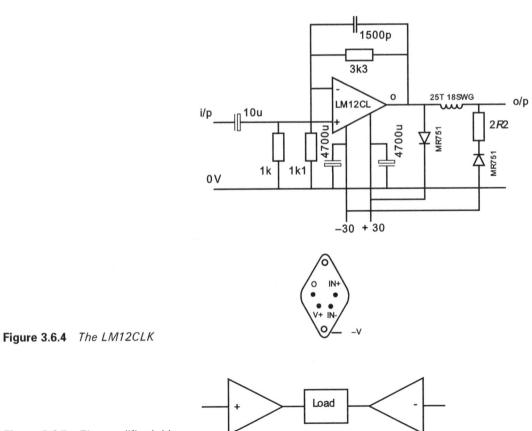

Figure 3.6.4 *The LM12CLK*

Figure 3.6.5 *The amplifier bridge connection*

Power amplifiers

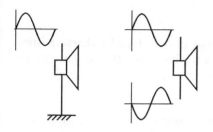

Figure 3.6.6 *Voltage waveforms as applied to the loudspeakers*

power out is given by $(V_{RMS})^2/R_L$, so there will be a useful increase in output power (Figure 3.6.6).

Wiring

These are high power, high gain devices, and some simple rules for circuit layout ought to be given:

(1) Keep the inputs physically separate from the outputs.
(2) Ensure the power supply leads are of adequate thickness for the power consumed by the amplifier.
(3) Bring all the circuit 0 V connections to one common 'star' point.
(4) All of the integrated circuits need ripple suppression capacitors on the supply lines. Ensure these are mounted as closely as possible to the power supply pins as is possible.
(5) Ensure that adequate heatsinking is provided.

Heatsinking

Devices which are not 100% efficient get hot. The inefficiency of the class of operation of the integrated circuit or output transistor manifests itself as heating. Heatsinks are rated in °C per watt as explained on in Chapter 1 (page 26). The same principle extends to discrete transistors and power amplifier ICs in that they must be prevented from overheating. As a rule of thumb, do not let the heat sink package exceed 70 °C because:

(1) an accidental touch could cause a burn;
(2) the life expectancy of the circuit could be compromised.

If the ambient temperature is 25 °C, then the heatsink must be chosen to keep the power component adequately cool.

Example 3.6.1

A 30 W Class B amplifier could have a maximum efficiency of 78.5%. In practice, it is likely to be lower than this, and to work on a figure of, say, 50% would be much more realistic. As the output power is reduced:

(i) the efficiency falls;
(ii) less power is demanded of the output device.

If 50% efficiency is assumed, with an ambient temperature of 25 °C, then to deliver 30 W requires 30 W dissipation within the amplifier and so:

Minimum heat sink rating $= (70 - 25)/30 = 1.5°$ C/W

This at best is only a rough estimate. Factors which will need to be taken into account could be case design; air flow; expected ambient temperatures; and diversity of output signal (how loud and for how long).

Distortion

All power devices have limitations in their **transfer characteristics**. That is, the relationship between the input and the output which it produces. In an ideal world, the current out I_o would be given by:

$$I_o = A \times V_i$$

In practice, the output for real transistors is a relationship more like:

$$I_o = A_0 + A_1 V_i + A_2 V_i^2 + A_3 V_i^3 + A_4 V_i^4 + \ldots$$

where V_i at its simplest, is a sine wave of the form:

$$V_i = V \sin \omega t$$

Successful manipulation of the terms which will follow from these equations involve the use of the trigonometrical double angle formulae.

Mathematics in action

$$\sin(A \pm B) = \sin A \cos B \pm \cos A \sin B$$

$$\sin 2A = 2 \sin A \cos A$$

$$\cos(A \pm B) = \cos A \cos B \mp \sin A \sin B$$

$$\cos 2A = \cos^2 A - \sin^2 A$$

As $\cos^2 A + \sin^2 A = 1$, then:

$$\cos 2A = 1 - 2\sin^2 A$$

$$\cos 2A = 2\cos^2 A - 1$$

Rearrangement would give:

$$\sin^2 A = \tfrac{1}{2} - \tfrac{1}{2} \cos 2A$$

$$\cos^2 A = \tfrac{1}{2} + \tfrac{1}{2} \cos 2A$$

$$\sin A \cos B = \tfrac{1}{2} \sin(A+B) - \tfrac{1}{2} \sin(A-B)$$

Applying these relationships to our non-ideal, but realistic, transfer characteristic:

$$V_i^2 = V^2 \sin^2 \omega t = \tfrac{1}{2}(1 - 2\cos 2\omega t) V^2$$

$$V_i^3 = V^3 \sin^3 \omega t = \tfrac{1}{2} \sin \omega t (1 - \cos 2\omega t) V^3$$

$$= \tfrac{1}{2}(\sin \omega t - \sin \omega t \cos 2\omega t) V^3$$

$$= \tfrac{1}{4}(3 \sin \omega t - \sin 3\omega t) V^3$$

If these are put into the transfer characteristic, and like terms in $\sin \omega t$, $\sin 3\omega t$, $\sin 5\omega t$, etc. collected together, we end up with:

$$I_o = \left(A_0 + \frac{A_2 V^2}{2}\right) + \left(A_1 V + \frac{3A_3}{4} V^3\right) \sin \omega t$$

$$- \frac{A_2 V^2}{2} \cos 2\omega t - \frac{A_3 V^3}{4} \sin 3\omega t$$

There is an ascending series in multiples of the harmonics of the fundamentals. The above equation could be simplified to:

$$I_o = a + b \sin \omega t - c \cos 2\omega t - d \sin 3\omega t + e \cos 4\omega t$$
$$+ f \sin 5\omega t - g \cos 6\omega t \ldots$$

The introduction of these extra frequencies at $2\omega t$ and $3\omega t$ give rise to the term **harmonic distortion**. Summing all the terms with unwanted frequencies produces the frequently quoted expression **total harmonic distortion** or THD.

If the amplitude of the second harmonic is D_2, the third D_3, the fourth D_4, etc., then the THD is given by the expression:

$$\text{THD} = \sqrt{D_2^2 + D_3^2 + D_4^2 + \ldots}$$

Problems 3.6.1

(1) Find the total harmonic distortion when the significant distortion components were found to be:

  D2 = 10%   D3 = 2%   D4 = 0.5%   D5 = 0.01%

(2) If only the odd harmonics were present, what would now be the total harmonic distortion?

(3) What is the total harmonic distortion caused by the presence of:

  D2 = 8%   D3 = 1.5%   D4 = 0.3%?

Class B harmonic distortion

In Class B, there are two output devices Q1 and Q2 which are responsible for passing the current for positive and negative half cycles of the output. When Q1 conducts, current flows into the load. When Q2 conducts, current flows out of the load. This was shown diagrammatically in Figure 3.4.1. The wave form fundamental is $(-\sin \omega t)$, so that the current sunk by Q2 is given by:

$$I_o = A_0 - A_1 \sin \omega t + A_2 \sin^2 \omega t - A_3 \sin^3 \omega t + A_4 \sin^4 \omega t - \ldots$$

which simplifies to:

$$I_o = -a + b \sin \omega t + c \cos 2\omega t - d \sin 3\omega t - e \cos 4\omega t$$
$$+ f \sin 5\omega t + g \cos 6\omega t \ldots$$

The inference is that with perfectly matched push–pull output transistors, the even harmonics will cancel, leaving only the odd harmonics.

It is an interesting fact that higher order harmonics are more disturbing to listen to than the lower order harmonics. Part of the reason for this is that the second harmonic is a doubling in frequency, which musically represents the octave above the fundamental. That is not disruptive to listen to. The fourth harmonic is a doubling of the second, which is the second octave above the fundamental. The higher odd harmonics, however, are not in any 'natural' musical relationship to the fundamental and so are quite intrusive. This is not to say that you must tolerate harmonic distortion. A lot of effort goes into audio amplifier design to eliminate harmonic distortion, and the figure of merit which is used is always the THD figure. There are other types of distortion which can occur in amplifiers, and these are of a lower amplitude than THD. One of these is the **intermodulation distortion**.

Intermodulation distortion

Intermodulation distortion occurs as a by-product of the non-linear input–output relationship of the amplifying devices. If two pure sine wave frequencies are input to the amplifier, the output will contain not only those frequencies, but also the sum and difference of these frequencies at a very low level. For example, if 400 Hz and 300 Hz were input to an amplifier with significant intermodulation distortion, the output would be 400 Hz, 300 Hz, plus 100 Hz (which is $f_1 - f_2$) and 700 Hz (which is $f_1 + f_2$).

Problems 3.6.2

(1) What would be the intermodulation products of an amplifier subjected to 200 Hz and 1000 Hz input signals?
(2) An amplifier has three signals of 900 Hz, 1200 Hz and 1300 Hz as inputs. What intermodulation products would you expect to be generated?

4 Feedback (2)

Summary

This chapter advances the ideas brought out in Chapter 2 (Feedback 1). In this chapter we will look at the effect of feedback on the input and output impedance of amplifiers. We will look at different ways in which feedback can be derived from the amplifier output and then applied to the amplifier input.

4.1 Amplifier classification

Amplifiers can be classified in one of four ways according to their input and output parameters:

(1) Voltage $A_v = \dfrac{v_O}{v_{IN}}$ e.g. op amp

(2) Current $A_i = \dfrac{i_O}{i_{IN}}$ e.g. BJT

(3) Transconductance $G_t = \dfrac{i_O}{v_{IN}}$ e.g. FET

(4) Transresistance $G_r = \dfrac{v_O}{i_{IN}}$ e.g. Norton op amp

Feedback itself can be:

> **derived** as a **voltage** or **current** and
> **applied** in **series** or **parallel** with the input signal.

Figure 4.1.1 shows these four possibilities schematically. In each case the feedback term β is not used. In practice, of course, it would modify the proportion of the term being fed back.

There is a gulf between feedback theory and its application, so this chapter will also consider some aspects of transistor amplifier design.

There now follows a set of four feedback circuits. These will be:

- Section 4.2 Series current (voltage current)
- Section 4.3 Series voltage (voltage voltage)

Figure 4.1.1 *Deriving and applying feedback*

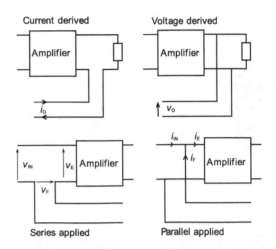

- Section 4.4 Shunt current (current current)
- Section 4.5 Shunt voltage (current voltage)

For each of these, an expression will be derived for gain, input impedance and output impedance. From these, you will be able to see that:

- current derived feedback **increases** the output impedance;
- voltage derived feedback **decreases** the output impedance;
- series applied (voltage) feedback **increases** the input impedance;
- parallel applied (current) feedback **decreases** the input impedance.

The trick is to identify from a given circuit diagram which type of feedback is which.

4.2 Series current

The feedback is proportional to the output current and applied in series with the input as shown in Figure 4.2.1.

The three basic equations relating the main amplifier, input signal mixing and feedback amplifier are:

$$A = \frac{i_O}{v_E} \quad v_E \times A = i_O \quad v_F = \beta \times i_O$$

The gain term for this transconductance amplifier is i_O/v_{IN} and the gain can be derived as:

$$v_{IN} - v_F = v_E$$

$$v_{IN} - \beta i_O = \frac{i_O}{A}$$

$$v_{IN} = i_O \left(\frac{1}{A} + \beta \right)$$

$$\frac{i_O}{v_{IN}} = \frac{A}{(1 + A\beta)}$$

As expected, the gain is reduced by the factor of $(1 + A\beta)$.

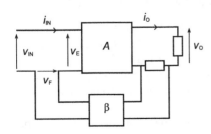

Figure 4.2.1 *Series current feedback*

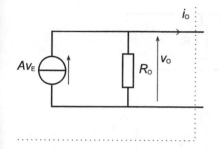

Figure 4.2.2 *The simplified amplifier output stage*

Input impedance

$$Z_{IN} = \frac{v_{IN}}{i_{IN}}$$

$$v_{IN} = v_E + v_F$$
$$= v_E + \beta i_O$$
$$= v_E + A\beta v_E$$

$$Z_{IN} = \frac{i_{IN} R_{IN} + A\beta i_{IN} R_{IN}}{i_{IN}}$$

$$Z_{IN} = R_{IN}(1 + A\beta)$$

Output impedance

$$Z_O = \frac{v_O}{i_O}$$

When $i_{IN} = 0$ then $v_{IN} = 0$, and then $-v_F = v_E$

$$Av_E = i_O - \frac{v_O}{R_O}$$

$$-Av_F = i_O - \frac{v_O}{R_O}$$

$$-A\beta i_O = -\frac{v_O}{R_O} + i_O$$

$$\frac{v_O}{i_O} = R_O(1 + A\beta) = R'_O$$

As stated, the input resistance is increased by a factor $(1 + A\beta)$ and the output resistance is increased by the same factor.

An undecoupled emitter resistor ... series current feedback

As an example, consider the circuit of a Class A BJT signal amplifier with the emitter resistor removed (Figure 4.2.3). The feedback is generated from the output current I_E and the voltage it generates when passing through R_E causes a voltage drop V_E which is applied in series with any input voltage.

In a.c. terms, v_I is approximately equal to the voltage on the emitter v_E, although there is a d.c. separation of 0.6 V between them due to the V_{BE} bias voltage.

$$v_E \approx v_I$$
$$v_E = i_E \times R_E$$
$$v_O = i_C \times R_C$$

So $$\frac{v_O}{v_E} = \frac{R_C}{R_E}$$

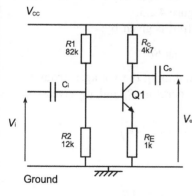

Figure 4.2.3 *A BJT amplifier*

As $v_I \approx v_E$:

$$\frac{v_O}{v_I} = \frac{R_C}{R_E}$$

Figure 4.2.4 Replacing the transistor with its a.c. equivalent circuit

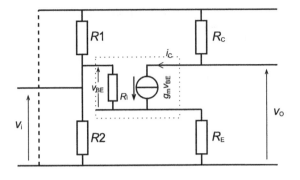

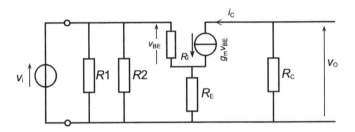

Figure 4.2.5 Redrawing the equivalent circuit

and as gain $\approx 1/\beta$:

$$\beta = \frac{R_E}{R_C}$$

The gain is dependent solely on the external components rather than any transistor parameters. This somewhat surprising result will be repeated when we analyse high gain operational amplifiers. It is, however, a useful feature, because one of the most difficult tasks is calculating the gain of an open loop transistor amplifier. The parameters h_{FE} and R_{IN} are so variable within any one batch of the same devices that precision is not possible.

An alternative proof looks into the small signal equivalent circuit. If the transistor is replaced with its (very) simplified a.c. equivalent, then the circuit of Figure 4.2.4 results. This only permits calculation of the a.c. response. The d.c. bias calculations form a separate exercise and would result in a different equivalent circuit.

Since we are only dealing with a.c. signals, the $+V_{CC}$ line can be connected to the 0 V line because they have a large ripple suppression capacitor between them which offers a very low impedance to a.c. This can be redrawn as in Figure 4.2.5.

The voltage drop across the output is given by:

$$v_O = -g_m v_{BE} R_C$$

While Kirchhoff's voltage law summing the voltages around the loop formed by the input signal generator, R_i and R_E gives:

$$v_I = v_{BE} + \left(\frac{v_{BE}}{R_i} + g_m v_{BE}\right)$$

Collecting together the v_{BE} terms:

$$v_I = v_{BE}\left(\frac{1}{R_E} + \frac{1}{R_i} + g_m\right)R_E$$

Eliminating v_{BE}:

$$\frac{v_O}{v_I} = \frac{-g_m R_C}{\left(\dfrac{1}{R_E} + \dfrac{1}{R_i} + g_m\right) R_E}$$

If $\dfrac{1}{R_E} + \dfrac{1}{R_i} < g_m$ then $\dfrac{v_O}{v_I} \approx -\dfrac{R_C}{R_E}$.

This confirms the approximate formula derived above.

D.c. design

It is one thing to evaluate the a.c. performance of the amplifier and another thing to design it in the first place. This short section shows the choices which are made when designing the Class A amplifier. The first step is to set up the correct bias conditions:

(1) V_{CC} The supply voltage is that which is available in the circuit. It can be a free choice for the designer or simply a matter of which supply is available.
(2) I_C The bias current I_C is normally between 0.1 mA and 10 mA. This value is arrived at because higher than 10 mA will cause heating effects in the transistor and less than 0.1 mA will require large values of bias resistor. The term **quiescent collector current** (I_{CQ}) is sometimes used to refer to the bias current which is present in the amplifier without an a.c. signal being applied to the circuit.
(3) V_E It has been found that for reasonable thermal stability, the emitter voltage V_E should be approximately one-tenth of V_{CC}.

The rest of the circuit design falls out from these three basic rules.

Let V_{CC} be 15 V and $I_C = 1.5$ mA. Then from rule 3 above, $V_E = 1.5$ V, which makes $R_E = 1$k.

The quiescent (no-signal) d.c. voltage of the collector should be half-way between that of V_E and V_{CC}. If it is made so, then the output can swing as far positive as it can negative. The voltage drop across R_C needs to be $(15 - 1.5)/2$. The current through it is set at 1.5 mA, so the nearest preferred value for R_C would be 4.7k.

When the value of V_E is 1.5 V, then V_B must be $1.5\,\text{V} + 0.6\,\text{V} = 2.1\,\text{V}$. The input resistors must then be in the ratio $R1:R2$ of $12.9:2.1$. The input resistance of the transistor is subject to some variability, so we cannot rely on its value. The h_{FE} of the transistor could reasonably be expected to be around 100, so the base bias current should be somewhere around 15 µA. (i.e. 1.5 mA/100). If we make the quiescent bias current through R1 and R2 very much larger than 15 µA, then we do not have to worry about the base bias current. Some trial and error work will be needed here, but with the values shown, the base bias current is around 159 µA and the ratio is approximately the correct value, given the preferred E12 range of resistors.

Without an emitter resistor the a.c. signal gain is shown to be -4.7 ('$-$' for an inverting amplifier, which this is). With a capacitor decoupling the emitter resistor, the gain would normally be in excess of -100. The next section shows how to modify the circuit to give a gain of any value (between those two limits) without affecting the d.c. bias.

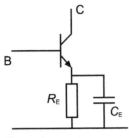

Figure 4.2.6 *The emitter decoupling capacitor*

Figure 4.2.6 shows the normal arrangement for the emitter resistor and its bypass capacitor.

The reactance of the capacitor would be given by:

$$X_C = \frac{1}{2\pi f C}$$

This reactance is in parallel with R_E, and as the frequency of the input signal falls, the reactance will eventually fall to the same number of ohms as the emitter resistor:

$$R_E = \left| \frac{1}{2\pi f C_E} \right|$$

$$f \approx \frac{1}{2\pi R_E C_E}$$

The parallel impedance falls to 0.707 of the mid-band value. Thus, the value of C sets the low frequency cut-off voltage as shown in Figure 4.2.7.

Example 4.2.1

In the example shown in Figure 4.2.3, calculate the low frequency roll-off when a 10 µF capacitor is connected across R_E.

$$f \approx \frac{1}{2\pi R_E C_E}$$

$$= \frac{1}{2 \times \pi \times 1k \times 10 \mu F}$$

$$= 16 \text{ Hz}$$

If the bias resistor is tapped (split into two resistors whose total value is still the same as the original) and only one of the resistors decoupled with C_E then it is possible to adjust the gain Figure 4.2.8).

The voltage gain will now be:

$$Av = \frac{R_C}{R_{E1}}$$

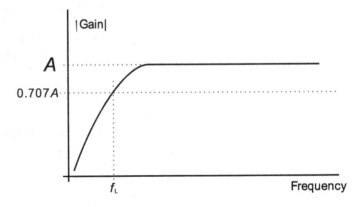

Figure 4.2.7 *The low frequency roll-off point*

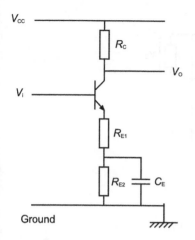

Figure 4.2.8 *A tapped emitter resistor*

Problems 4.2.1

(1) Design a BJT amplifier to operate from a 12 V power supply with a collector current of 5 mA. Adapt your design for an overall a.c. voltage gain of −20 and low frequency roll-off frequency of 20 Hz.
(2) Design an amplifier to operate from a 9 V battery supply with a collector current of 0.4 mA. It should have an a.c. voltage gain of −15 and low frequency roll-off of 25 Hz.
(3) Design a 5 V amplifier with a gain of −30 and low frequency roll-off of 10 Hz.

JFET design

The same principle applies to FET-based circuits. That is, if the source bypass capacitor is removed as in Figure 4.2.9:

- the a.c. voltage gain will be determined by the ratio R_D/R_S;
- the input impedance will be increased;
- the output impedance will be increased.

The circuit is not often used because this form of FET circuit has but a low a.c. voltage gain to start with (−5 to −15 typically). Reduction of gain with feedback reduces the usability of the circuit unless some of its other properties are required such as increase of input and output impedances. JFETs inherently have a high input impedance anyway.

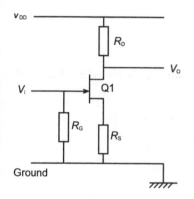

Figure 4.2.9 *FET amplifier with undecoupled source resistor*

4.3 Series voltage

The feedback is proportional to the output voltage, and is applied in series to the input (Figure 4.3.1). The basic equations are:

$$v_{IN} - v_F = v_E \quad \beta v_O = v_F \quad v_O = Av_E$$

From this:

$$V_{IN} - \beta v_O = \frac{v_O}{A}$$

$$v_{IN} = v_O\left(\frac{1}{A} + \beta\right)$$

$$v_{IN} = v_O\left(\frac{1 + A\beta}{A}\right)$$

$$\frac{v_O}{v_{IN}} = \frac{A}{1 + A\beta}$$

The gain is reduced by the factor of $(1 + \beta)$.

Figure 4.3.1 *The series voltage amplifier*

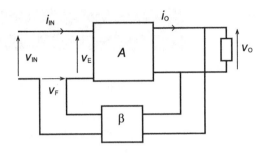

Input impedance

$$v_{IN} = v_E + v_F$$
$$= I_E R_E + \beta v_O$$
$$= I_E R_E + A\beta v_E$$
$$= I_E R_E + A\beta I_E R_E$$
$$\frac{v_S}{I_E} = R'_{IN} = R_E(1 + A\beta)$$

Output impedance

For simplicity, assume that $v_{IN} = 0$ and apply an external signal to the output of value v_O. Then:

$$-v_F = v_E \quad \text{and} \quad v_F = \beta v_O$$
$$v_O = I_O R_O + A v_E$$
$$= I_O R_O - A\beta v_O$$
$$v_O(1 + A\beta) = I_O R_O$$
$$\frac{v_O}{I_O} = R'_O = \frac{R_O}{1 + A\beta}$$

It can be seen that the output impedance is decreased, while the input impedance is increased.

Figure 4.3.2 shows two simple applications of this mode of feedback. In each case, the voltage gain is just less than unity. These circuits are used for their voltage buffering abilities rather than their voltage gain properties.

Figure 4.3.3 shows the a.c. small signal equivalent circuit for the JFET amplifier. The analysis for the JFET source follower is:

$$v_{IN} = v_{GS} + v_F$$
$$v_O = g_m v_{GS} \times R_S \| r_{DS}$$
$$v_{IN} = v_O + \frac{v_O}{g_m R_S \| r_{DS}}$$
$$v_{IN} = v_O \left(\frac{g_m R_S \| r_{DS} + 1}{g_m R_S \| r_{DS}} \right)$$
$$A_v = \frac{v_O}{v_{IN}} = \frac{g_m R_S \| r_{DS}}{1 + g_m R_S r_{DS}} \approx 1$$

R_O is set by the parallel combination of R_S and r_{DS}.

Figure 4.3.2 *Emitter and source follower circuits*

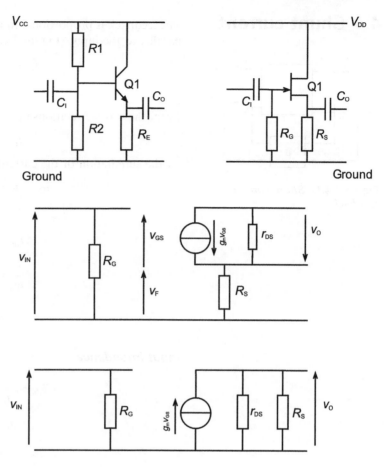

Figure 4.3.3 *The field effect transistor source follower*

The long tailed pair

The long tailed pair is another form of series voltage feedback (Figure 4.3.4).

The voltage feedback factor is:

$$\frac{1}{1+9.1} \approx \frac{1}{10}$$

giving a voltage gain of 10 in this circuit.

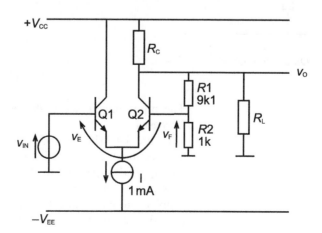

Figure 4.3.4 *The long tailed pair*

4.4 Shunt current

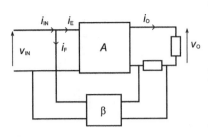

Figure 4.4.1 Shunt current feedback

The feedback is proportional to the output current and is applied in parallel to the input (Figure 4.4.1):

$$i_E \times A = i_O$$

$$i_F = \beta \times i_O$$

From these basic relationships:

$$i_F = A\beta i_E$$

At the summation of the current:

$$i_{IN} - i_F = i_E$$

$$i_{IN} - A\beta i_E = i_E$$

$$i_{IN} = i_E(1 + A\beta)$$

$$\frac{i_O}{i_{IN}} = A' = \frac{A}{1 + A\beta}$$

Input impedance

$$R_{IN} = \frac{v_{IN}}{i_E}$$

$$i_{IN} = i_F + i_E$$

$$i_{IN} = \beta i_O + i_E$$

$$i_{IN} = i_E(1 + A\beta)$$

$$\frac{v_{IN}}{i_{IN}} = R'_{IN} = \frac{R_{IN}}{1 + A\beta}$$

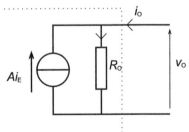

Figure 4.4.2 Output impedance

Output impedance
When $i_{IN} = 0$:

$$i_E = -i_F$$

$$i_F = \beta i_O$$

$$i_O - A i_E = \frac{v_O}{R_O}$$

$$(1 + A\beta)R_O = \frac{v_O}{i_O}$$

The input impedance is **decreased**, and the output impedance is **increased**.

Figure 4.4.3 shows a BJT amplifier with shunt current feedback. The feedback is proportional to the output current (through R_E) and applied in parallel to the base of Q1 by injecting the feedback current i_F. This can be simplified into Figure 4.4.4. This seems simpler, but may take a few moments to work through. R1 not only provides feedback, but also the base bias for Q1.

Figure 4.4.3 An amplifier with shunt current feedback

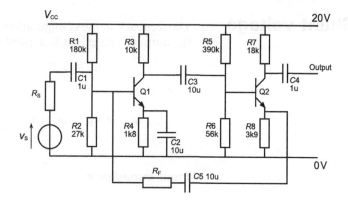

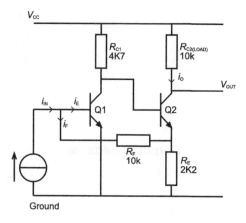

Figure 4.4.4

The design equations can be shown to be:

$$I_{C2} \times R_{E2} = I_{B1} \times R_F - 0.6$$

$$I_{C2} \times R_{E2} = \frac{I_{C1}}{h_{FE}} \times R_F - 0.6 \tag{1}$$

$$V_{CC} = I_{C1} R_{C1} + 0.6 + I_{C2} R_{E2} \tag{2}$$

Between simultaneous equations (1) and (2), the values of I_{C1} and I_{C2} can be determined if a value of h_{FE} is assumed. (150 is as good as any value for a rough estimate.)

Problems 4.4.1

(1) Determine I_{C1} and I_{C2} in Figure 4.4.4. (Assume $V_{CC} = 20$ V and that $h_{FE} = 150$.)
(2) Design a feedback amplifier of the style of Figure 4.4.4 where:

$$V_{CC} = 12 \text{ V}$$
$$I_{C1} = 0.5 \text{ mA}$$
$$I_{C2} = 2 \text{ mA}$$
$$V_{out} \text{ (d.c.)} = 7 \text{ V}$$

4.5 Shunt voltage

The feedback in the configuration of Figure 4.5.1 is proportional to the output voltage and is applied in parallel to the input:

$$i_{IN} - i_F = i_E \quad i_F = \beta v_O \quad v_O = A i_E$$

From these:

$$i_F = A\beta i_E \quad \text{and} \quad A' = \frac{v_O}{i_{IN}} = \frac{A i_E}{i_E + A\beta i_E} = \frac{A}{1 + A\beta}$$

Input impedance

$$i_{IN} - i_F = i_E$$
$$i_{IN} = i_E + \beta v_O$$
$$i_{IN} = i_E + A\beta i_E$$
$$i_{IN} = i_E(1 + A\beta)$$
$$R' = \frac{R}{1 + A\beta}$$

Output impedance

$$i_E = -i_F$$
$$i_F = \beta v_O$$
$$v_O = A i_E + i_O R_O$$
$$v_O = -A\beta v_O + i_O R_O$$
$$v_O + A\beta v_O = i_O R_O$$
$$R' = \frac{R_O}{1 + A\beta}$$

The shunt voltage configuration reduces both the input and output impedance.

A simple transistor version of the shunt voltage circuit is shown in Figure 4.5.2. This circuit reduces both the input and output impedance. The circuit can be simplified to one of two circuits, as shown in Figures 4.5.3 and 4.5.4.

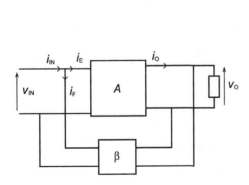

Figure 4.5.1 *Shunt voltage feedback*

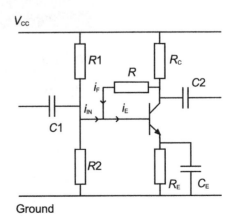

Figure 4.5.2 *A simple shunt voltage amplifier*

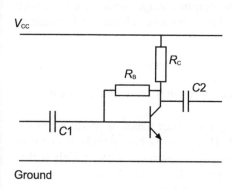

Figure 4.5.3 *A simpler example*

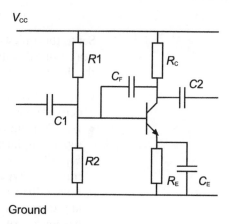

Figure 4.5.4 *Using a capacitor for feedback*

R_B in Figure 4.5.3 provides both d.c. bias and a.c. feedback. This simple-looking circuit provides a good basis for many an a.f. signal amplifier. The feedback resistor does limit the gain, but overall performance is useful. Unfortunately, the gain cannot be determined by the value of the external components as in earlier circuits. A knowledge of the transistor parameters is needed for precise gain figure prediction. Since any specific transistor type has a wide parameter spread, it would be misleading to generalize. However, an analysis and guide for design follows shortly.

Capacitor C_F in Figure 4.5.4 provides frequency-dependent feedback. As the frequency increases, so the impedance of the capacitor falls. At high frequencies, the capacitor will provide a low impedance path between the collector and the base – i.e. between the input and the output.

This reduces the gain at high frequencies. The roll-off frequency is not a simple function of the external components, but also of the internal transistor characteristics. This circuit is often used to limit the frequency response of amplifier gain stages. Figure 4.5.5 shows the effect of increasing C_F on the frequency response of the amplifier.

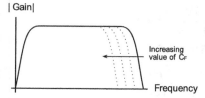

Figure 4.5.5 *Amplifier frequency response*

Designing a simple shunt voltage feedback amplifier

If we are to build a circuit as in Figure 4.5.6, the ideal value of quiescent collector voltage would be half of the supply voltage:

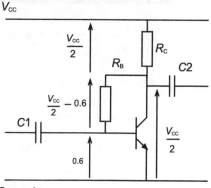

Figure 4.5.6 *Circuit for analysis*

$$V_C = \tfrac{1}{2} V_{CC}$$

Set to this quiescent value, the output voltage can have a maximum peak-to-peak value of $\tfrac{1}{2} V_{CC}$. There are other choices to be made in the design of the circuit:

- V_{CC} This depends on the local supplies (either battery or derived) which are available.
- I_C Conventionally, this lies between 0.1 mA and 10 mA in most signal amplifiers.
- h_{FE} To some extent, this is a choice which is made when the transistor type is selected. However, as has already been mentioned, the value of forward current gain h_{FE} varies widely within any transistor sample. For example, the BC549 signal transistor has an h_{FE} of typically between 200 and 800.

Example 4.5.1

Design an amplifier so that $V_{CC} = 10\,\text{V}$, $I_C = 1\,\text{mA}$ and $h_{FE} = 100$.

In practice, when choosing a transistor, it will have a different value of h_{FE} than this, so I_C and V_{RC} will not be those initially used in the calculation.

$$V_{RC} = 5\,\text{V (ideally)}$$

$$I_{RC} \approx I_C = 1\,\text{mA}$$

$$R_C = \frac{5\,\text{V}}{1\,\text{mA}} = 5\,\text{k}\Omega$$

$$V_{RB} = 5\,\text{V} - 0.6\,\text{V}$$

$$I_{RB} = \frac{I_C}{h_{FE}} = \frac{1\,\text{mA}}{100}$$

$$R_B = \frac{4.4\,\text{V}}{10\,\mu\text{A}} = 440\,\text{k}\Omega$$

Choose $R_C = 5\text{k}6$ $R_B = 470\text{k}$ or $R_C = 4\text{k}7$ $R_B = 390\text{k}$ to the nearest preferred values.

Problems 4.5.1

(1) Design the circuit of Figure 4.5.3 to operate from an 8 V power supply when $I_C = 0.8\,\text{mA}$. Assume that $h_{FE} = 150$.

(2) If the circuit of Figure 4.5.3 has $I_C = 1.2\,\text{mA}$, calculate the bias components for $V_{CC} = 14\,\text{V}$ and $h_{FE} = 200$.

(3) Using the values of R_C and R_B calculated for the circuit of question 2, if the transistor was selected from a type which actually has an h_{FE} range of 100 to 300:

  (a) calculate V_C when $h_{FE} = 100$;
  (b) calculate V_C when $h_{FE} = 300$.

(You will find that for the different values of h_{FE}, the collector and base currents will alter from those initially assumed in question 2. This sort of cross-checking is often needed with transistor circuits to ensure that they will work as designed irrespective of the value of h_{FE}.)

(4) Design a transistor amplifier of the type of Figure 4.5.3 which is to operate from a 12 V power supply. The transistor is to be a BC337A with h_{FE} of 100 to 400. Assume a collector current of 1.5 mA.

4.6 Feedback problems

(1) In the circuit of Figure 4.6.1:
  (a) What class of feedback is this?
  (b) Which components control the feedback ratio β?
  (c) What is the value of β and hence the approximate closed loop gain?
  (d) Calculate the d.c. static circuit values:
  (i) V_{B1}
  (ii) V_{E1}
  (iii) I_{R4}
  (iv) V_{R3}
  (v) I_{R3}
  (vi) I_{R5}
  (vii) V_O

(2) In the circuit of Figure 4.6.2:
  (a) Identify the feedback components in this circuit.
  (b) Which classification of feedback is it?
  (c) What is the feedback factor β?
  (d) What is the approximate gain of the circuit?

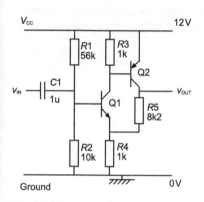

Figure 4.6.1 Feedback circuit

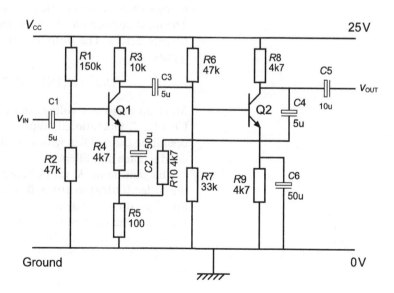

Figure 4.6.2 Another BJT feedback circuit

5 Operational amplifiers (1)

Summary

This is the first of two chapters covering operational amplifiers. By the end of this chapter, you should be able to understand the basic concepts of these important devices and be able to design them into circuits with due regard to appropriate selection for their parameters. These will be limited to common mode rejection ratio, slew rate, input offset voltage and current and gain bandwidth product.

5.1 Introduction

An operational amplifier (or op amp) is a complex circuit of resistors, capacitors and transistors integrated as one circuit onto a single piece of silicon. Hence the expression **integrated circuit** (IC). The most common form of op amp has two inputs and one output (Figure 5.1.1). The output is the amplified difference of the two inputs:

$$\text{Out} = K(\text{IN1} - \text{IN2}) \quad \text{where} \quad K = \text{gain}$$

Most op amps are voltage devices so that $V_o = K(V1 - V2)$. (In Chapter 7, Operational amplifiers 2, we will look at other types of operational amplifier which output current – these are the transconductance amplifiers.)

The gain figure K varies between device types but is typically of the order 100 000 to 10 000 000 (see Table 5.1.1 for a comparison). So, if an op amp has a gain of 100 000 ($= 10^5$ or 100 dB) then in theory, if the difference between $+V_{in}$ and $-V_{in}$ were only 10 µV, then the output would be 1 V.

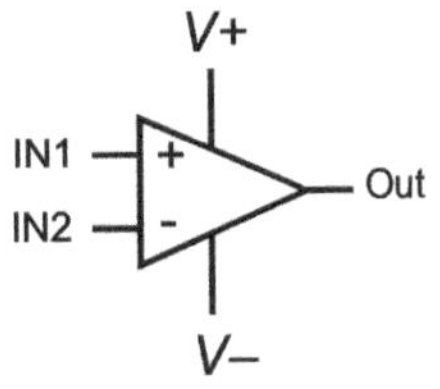

The inputs $+V$ and $-V$ are the power supply inputs for the integrated circuit. These could be bipolar (dual rail), say +15 V and −15 V or +12 V and −12 V, or single rail, say +15 V and 0 V. Table 5.1.1 also shows the range of a very few common devices. When choosing an op amp, care has to be exercised to ensure that the device has an acceptable output swing for the application.

Figure 5.1.1 *The operational amplifier*

Operational amplifiers (1) 103

Table 5.1.1 *Operational amplifiers compared*

Op amp device	Bipolar or field effect	Gain (dB)	Supply (V)	Output swing from 15 V supplies	Slew rate (V/μs)	GBW (Hz)	CMRR (dB)	Input offset voltage	Input offset current
LM308	B	102	5–18	13	0.15	0.3	80	2 mV	1 nA
NE531	B	96	5–22	15	35	1	70	2 mV	40 nA
OP07	B	132	3–18	13	0.17	0.6	106	30 μV	3.8 nA
LM741	B	100	5–18	13	0.5	1.2	70	1 mV	20 nA
LF355	F	106	4–18	13	5	2.5	85	3 mV	20 pA
TL071	F	106	3–18	13.5	13	1	70	3 mV	50 pA
AD743	F	152	4.8–18	13.6	2.8	4.5	95	250 μV	40 pA

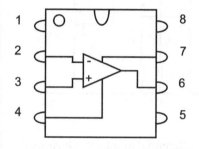

Figure 5.1.2 *DIL outline package*

Packages

The most common package for the op amp is the 8 pin dual-in-line (8 pin DIL). The connections shown in Figure 5.1.2 have developed into an industry standard.

The purpose of the remaining pins 1, 2 and 8 depend on the exact type of op amp. Other packages are shown in Figure 5.1.3.

What's inside

Figure 5.1.4 shows the internal construction of an op amp. Although it looks daunting, you should already be able to recognize most of the internal structures.

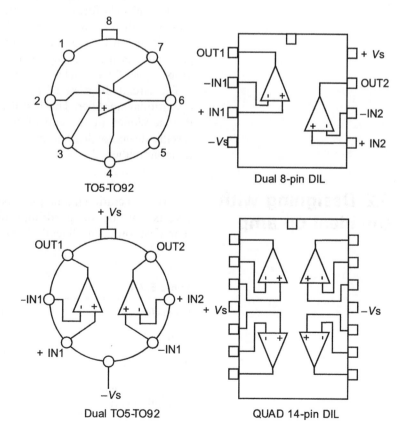

Figure 5.1.3 *Other outline packages*

Figure 5.1.4 *The internal schematic of an operational amplifier*

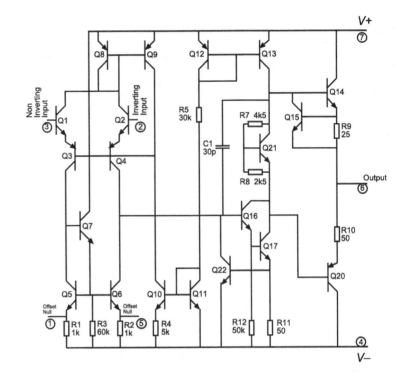

Transistors Q1 and Q2 make a differential amplifier which works with Q3 and Q4 to form a long tailed pair. Q5, Q6 and Q7 are the loads of the long tailed pair. Transistor pairs Q8/Q9 and Q10/Q11 are two constructions called current mirrors. A change of collector current in Q8 causes a corresponding change in the collector current of Q9.

Q14 and Q20 form a conventional Class AB output stage with crossover distortion eliminated by the V_{BE} multiplier Q21. The stage is driven by the Darlington pair Q16/Q17. Output overload protection is provided by Q15 and Q22 which sense the voltage drop across $R9$ and $R11$ as discussed in Chapter 1.

The circuit diagram of Figure 5.1.4 is typical of a '741' type operational amplifier. This is still a useful, if aged, device which was first developed at the end of the 1960s. The differences between specific devices will be discussed more fully in Chapter 7 (Operational amplifiers 2).

5.2 Designing with the ideal op amp

A perfect operational amplifier does not exist, but useful functional circuits can be designed and operated taking the crudest of simplifications. The 'ideal' characteristics are as shown in Table 5.2.1.

Table 5.2.1

Parameter	'Ideal'	'Realistic'
Gain	Infinite	10^5–10^6
Input impedance	Infinite	10^6–$10^{12}\,\Omega$
Bandwidth	Infinite	$\sim 1\,\text{MHz}$
Noise	Zero	$\sim 50\,\text{nV}/\sqrt{\text{Hz}}$
Input offset voltage	Zero	$\sim 1\,\text{mV}$
Input offset current	Zero	$\sim 1\,\text{nA}$

Operational amplifiers (1)

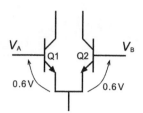

Figure 5.2.1 *The simplified input stage*

There are **five** basic configurations to study and understand. First, there is the basic concept of an operational amplifier **virtual earth** to explain. The input stage could be simplified to that of a long tailed pair as in Figure 5.2.1.

Each transistor has a base–emitter voltage of 0.6 V, so both base voltages will be very close in value, i.e.:

$$V_A \approx V_B$$

It is the minute (µV) difference between these which is amplified.

The circuits which follow adopt the common convention that the power supply connections are omitted for clarity. They have to be connected, of course, but a separate diagram is often issued to show how the power supplies are connected to all of the operational amplifiers.

Circuit 1 The buffer amplifier

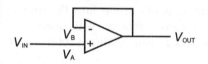

Figure 5.2.2 *The buffer amplifier*

$$V_A = V_B$$
$$V_{IN} = V_B$$
$$V_B = V_O$$

Hence $V_{IN} = V_O$.

We have designed a circuit for which the output is the same as the input. So what? Well, this circuit is most frequently used as a buffer where the high (infinite?) input impedance will not act as a drain on the source of the signals. For example, a simple photodetector could be made using a photodiode as shown in Figure 5.2.3.

Incident light on the photodiode increases its reverse leakage current, but even so it is still only in the nA to µA region. A high input impedance ensures that the tiny amount of current leaking through the diode is not used to bias the input stage of the circuit connected to it.

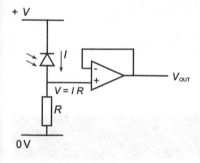

Figure 5.2.3 *Simpler photodetector using a diode*

Circuit 2 The non-inverting amplifier

The circuits in Figure 5.2.4 are identical, but can be drawn either way around. $R1$ and $R2$ form a potential divider (can you identify

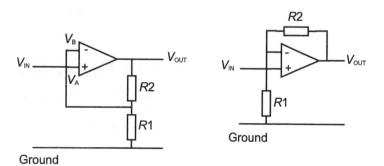

Figure 5.2.4 *The non-inverting amplifier*

the feedback classification?) so that:

$$\frac{V_B}{V_O} = \frac{R1}{R1+R2}$$

$$V_A = V_B \quad \text{and} \quad V_{IN} = V_A$$

so $$\frac{V_{IN}}{V_O} = \frac{R1}{R1+R2}$$

or, more simply:

$$\text{Gain} = Av = \frac{V_{OUT}}{V_{IN}} = \frac{R1+R2}{R1}$$

The gain is controlled solely by the feedback components as discussed in Chapter 4.

There are two points to note:

(1) Try to keep resistor values between 1 kΩ and 1 MΩ. Less than 1 kΩ can cause problems with the amplifier inputs trying to source or sink significant amounts of current into the source resistors; more than 1 MΩ can cause problems with noise.
(2) Negative feedback – the feedback resistor is always returned to the negative signal input.

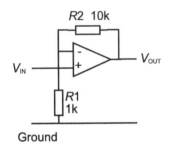

Figure 5.2.5

Example 5.2.1

Find the gain of the circuit of Figure 5.2.5.

$$\text{Gain} = \frac{10k + 1k}{1k} = 11$$

Problem 5.2.1

In the circuit of Figure 5.2.6:

(a) Calculate the gain of circuit (a).
(b) Calculate the output voltage in the circuit (b).
(c) Calculate the value of R1 which will give the output voltage shown in circuit (c).
(d) Calculate the value of R2 which will give the output voltage shown in circuit (d).

Circuit 3 The inverting amplifier

In Figure 5.2.7, V_A is connected to ground, and since V_B is at the same potential as V_A, V_B is called a **virtual earth**.

The current through $R1$ is given by I_{R1} where:

$$I_{R1} = \frac{V_{IN} - V_B}{R1}$$

Operational amplifiers (1) 107

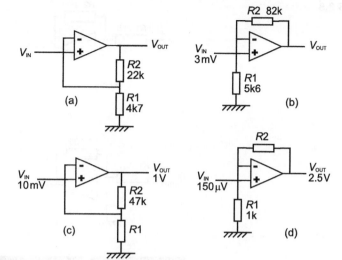

Figure 5.2.6

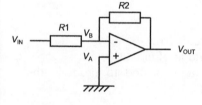

Figure 5.2.7 *The inverting amplifier*

Since $V_B \approx V_A = 0\,\text{V}$ then:

$$I_{R1} = \frac{V_{IN}}{R1}$$

The input impedance is very large, so not much of this current enters the op amp – most passes through $R2$.

The current through $R2$ is given by I_{R2} where:

$$I = \frac{V_B - V_{OUT}}{R2}$$

$$V_B \approx V_A = 0\,\text{V}$$

so $$I_{R2} = -\frac{V_{OUT}}{R2}$$

$$I_{R1} \approx I_{R2}, \quad \text{so:}$$

$$\frac{V_{IN}}{R1} = -\frac{V_{OUT}}{R2}$$

which gives the gain as:

$$\frac{V_{OUT}}{V_{IN}} = -\frac{R2}{R1}$$

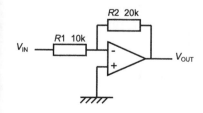

Figure 5.2.8

Example 5.2.2

Calculate the gain of the amplifier of the circuit in Figure 5.2.8.

The gain of the amplifier is given by:

$$\frac{V_{OUT}}{V_{IN}} = -\frac{R2}{R1} = -\frac{20k}{10k} = -2$$

Figure 5.2.9

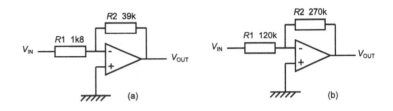

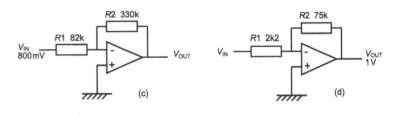

Problem 5.2.2

In the circuit of Figure 5.2.9:

(a) Calculate the gain of circuit (a).
(b) Calculate the gain of circuit (b).
(c) Calculate the output voltage of circuit (c).
(d) Calculate the input voltage needed to produce the output voltage of circuit (d).

Circuit 4 The mixer or adder

The circuit in Figure 5.2.10 sums the voltages at its inputs. In this circuit, the current through the input resistors $R1$, $R2$ and $R3$ sum at the $-$IN point:

$$I_{IN} = \frac{V1}{R1} + \frac{V2}{R2} + \frac{V3}{R3}$$

This current passes through the feedback resistor $R4$ and the output voltage is given by:

$$\frac{0 - V_{OUT}}{R4} = I_F$$

$I1 + I2 + I3 = I_F$ so:

$$\frac{V1}{R1} + \frac{V2}{R2} + \frac{V3}{R3} = -\frac{V_{OUT}}{R4}$$

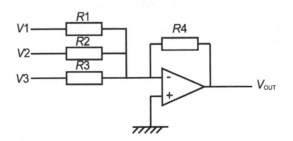

Figure 5.2.10 *The mixer*

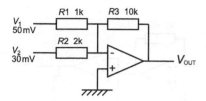

Figure 5.2.11 *Mixer for Example 5.2.3*

There is no 'gain' term as such. The smaller the input resistor, the larger the input current, and this gives a greater 'weighting' or importance of a particular input.

Example 5.2.3

In the circuit of Figure 5.2.11, calculate the expected output.

First notice the slightly different construction. Electrically, it is identical to the previous problem (apart from the fact that there are only two input resistors, that is). This economy of drawing is quite common and should be recognized as easily as the first method.

$$\frac{V1}{R1} + \frac{V2}{R2} = -\frac{V_{OUT}}{R3}$$

$$\frac{0.05}{1000} + \frac{0.03}{2000} = -\frac{V_{OUT}}{10\,000}$$

from which:

$$V_{OUT} = -65\,\text{mV}$$

Problems 5.2.3

(1) In the circuits of Figure 5.2.12, find the output voltage.
(2) In the circuit of Figure 5.2.13, which involves several of the amplifier types studied so far, calculate the output voltage.
(3) Calculate the output voltage of the circuit in Figure 5.2.14.

Circuit 5 The differential amplifier or subtractor

As its name implies, this circuit will output a voltage proportional to the difference in the two inputs. The basic circuit diagram is shown in Figure 5.2.15.

By voltage division at the +IN input:

$$V_A = \frac{R4}{R3 + R4}V1$$

By current flow at the −IN input:

$$\frac{V2 - V_B}{R1} = \frac{V_B - V_O}{R2}$$

$$\frac{V2}{R1} - \frac{V_B}{R2} = \frac{V_B}{R2} - \frac{V_O}{R2}$$

$$\frac{V2}{R1} + \frac{V_O}{R1} = V_B\left(\frac{1}{R1} + \frac{1}{R2}\right) = V_B\left(\frac{R1 + R2}{R1R2}\right)$$

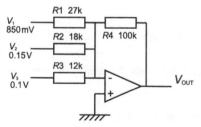

Figure 5.2.12

Figure 5.2.13

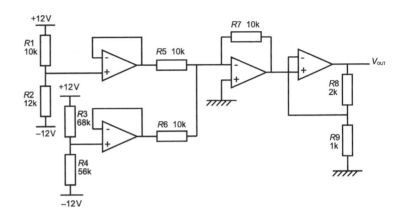

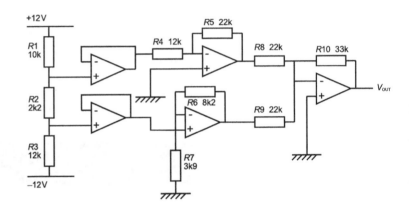

Figure 5.2.14

$V_A = V_B$ so:

$$\frac{V2}{R1} + \frac{V_O}{R2} = V1\frac{R4}{R3+R4}\left(\frac{R1+R2}{R1R2}\right)$$

$$V1\left[\frac{R4}{R3+R4}\left(\frac{R1+R2}{R1R2}\right)\right] - \frac{V2}{R1} = \frac{V_O}{R2}$$

In most applications, $R1 = R3$ and $R2 = R4$. Making this substitution:

$$V1\left(\frac{R2}{R1+R2} \times \frac{R1+R2}{R1R2}\right) - \frac{V2}{R1} = \frac{V_O}{R2}$$

$$\frac{V1}{R1} - \frac{V2}{R1} = \frac{V_O}{R2}$$

The final 'gain' term results in:

$$V_O = (V1 - V2)\frac{R2}{R1}$$

To get a gain of 10, you could use $R2 = R4 = 10k$ and $R1 = R3 = 1k$.

Operational amplifiers (1)

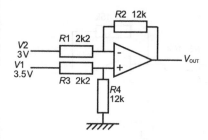

Figure 5.2.15 *The differential amplifier*

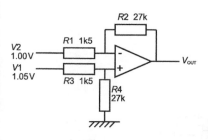

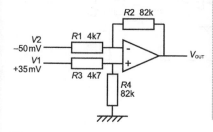

Figure 5.2.16

Example 5.2.4

In the circuit of Figure 5.2.15, calculate the expected output voltage.

$$V1 = 3.5\,\text{V}$$
$$V2 = 3.0\,\text{V}$$
$$R1 = R3 = 2.2\,\text{k}\Omega$$
$$R2 = R4 = 12\,\text{k}\Omega$$
$$V_{\text{OUT}} = (V1 - V2)\frac{R2}{R1}$$
$$= 0.5 \times \frac{12}{2.2} = 2.73\,\text{V}$$

Problems 5.2.4

(1) In the circuits of Figure 5.2.16, calculate the output voltages.
(2) In the circuit of Figure 5.2.17, calculate the output voltage.
(3) In the circuit of Figure 5.2.18, calculate the output voltage.

5.3 Other operational parameters

Bandwidth

Bandwidth is related to the range of frequencies over which the amplifier will operate (Figure 5.3.1). It could be specified in one of several ways:

(i) first breakpoint on the frequency response graph;
(ii) frequency at which the gain falls to unity;
(iii) full power bandwidth;
(iv) gain–bandwidth product.

Of the options, (ii) and (iv) are most often used.

Item (i) shown in the diagram is misleading to some extent. The op amp shown would appear to have a bandwidth of 10 Hz by this reckoning. Item (ii) gives a bandwidth of 10^6 Hz. Item (iii) is a possibility, but is always less than (ii) because of a factor called slew rate limiting (see below). Item (iv) is a useful factor, which given an ideal linear response will be the same as (ii).

Slew rate limiting

The **slew rate** is the maximum rate at which the output can follow a

Figure 5.2.17

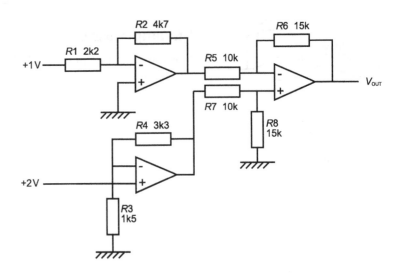

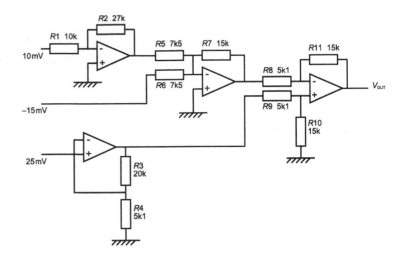

Figure 5.2.18

step input. Figure 5.3.2 shows how an output might be slew rate limited when the input is driven by a square wave.

The rate is quoted in volts per microsecond or V/μs:

Op amp	709	741	TL081	531
Slew rate (V/μs)	12	0.5	13	35

The output could slew positive at a different rate than when it slews negative because of different active component charge storage effects in the op amp itself.

When driving a sine wave into an op amp, the signal is characterized by the equation:

$$v = \hat{V} \sin \omega t$$

Slew rate limiting will limit the highest frequency signal to:

$$f = \frac{S}{2\pi \hat{V}}$$

Figure 5.3.1 *Operational amplifier frequency response*

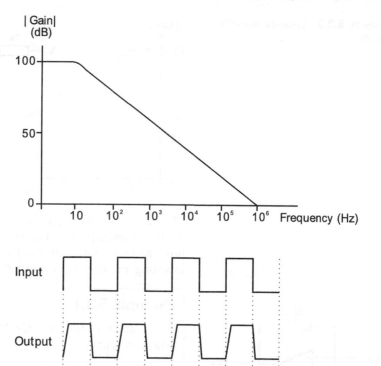

Figure 5.3.2 *Slew rate limiting*

Mathematics in action

When $v = \hat{V}\sin\omega t$ the rate of change is given by dv/dt:

$$\frac{dv}{dt} = \hat{V}\omega\cos\omega t$$

Now, $\cos\omega t$ is a function which can take a value of between 0 and 1. The maximum value of $\cos\omega t$ would be 1. Hence the fastest rate of change of the voltage is given by:

$$\frac{dv}{dt} = \hat{V}\omega (= \hat{V}2\pi f)\,\text{V/s}$$

If the slew rate is S, then the highest frequency which can be handled by the operational amplifier will be:

$$f = \frac{S}{2\pi \hat{V}}$$

One of the most useful measures of bandwidth is with the figure of merit called **gain–bandwidth product** or GBW.

The well known 741 has a gain–bandwidth product of approximately 106 Hz. This means that it would have:

- a gain of 1000 and a bandwidth of 1 kHz;
- a gain of 100 and a bandwidth of 10 kHz;
- a gain of 10 and a bandwidth of 100 kHz; etc.

Figure 5.3.3 *Gain–bandwidth product*

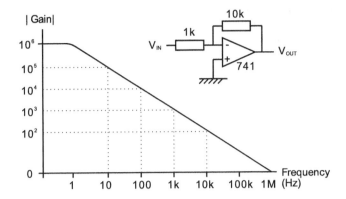

The gain is set by the resistive feedback components. The bandwidth is a consequence of manipulating the gain. Figure 5.3.3 shows how the bandwidth is increased as the gain is reduced. Inset is the inverting amplifier configuration.

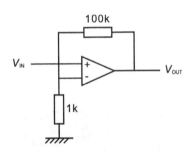

Figure 5.3.4

Example 5.3.1

An LF 355 op amp has a GBW product of 2.0 MHz and a slew rate of 5 V/µs.

(a) What would be its bandwidth if built into the circuit of Figure 5.3.4?
(b) When a 50 mV RMS signal is input, what is the maximum frequency which could be handled without slew rate limiting?

(a) The gain of the circuit is given by:

$$Av = \frac{100k + 1k}{1k}$$

$$Av = 101$$

$$GBW = Av \times BW$$

$$2\,MHz = 101 \times BW$$

$$BW = 19.8\,kHz$$

(b)

$$S = 5\,V/\mu s$$

$$\hat{V} = 50\,mV \times \sqrt{2} \times 101 = 7.14\,V$$

$$f = \frac{S}{2\pi \hat{V}}$$

$$= 111\,kHz$$

There seems to be some contradiction between the results of (a) and (b). The bandwidth would be limited to 19.8 kHz as presented by the gain–bandwidth calculation. Beyond this frequency, the gain rolls off and the output is not as large as $(101 \times V_{IN})$ would give you to believe. Above 111 kHz, the output is actually distorted for any signals as large as $7.14\,V_{pk}$. This means that there are two criteria to be observed:

(i) Will the gain–bandwidth product relationship allow you to get the expected bandwidth for the calculated gain?

Figure 5.3.5 *Slew rate limit for a sine wave*

(ii) At what amplitudes and frequencies will slew rate distortion start to affect the output?

Figure 5.3.5 shows the maximum slew rate which is acceptable for a sine wave. If the frequency is any higher than this, then distortion will result.

Multistage amplifiers

To keep the bandwidth at 20 kHz with a higher gain than 101, then more stages would be needed. For example, two such units cascaded would give a gain of $101 \times 101 = 10\,201$ while maintaining a bandwidth of 20 kHz.

Extending this, if the two cascaded amplifiers are set to have a gain of 10 and 20, the overall gain would be $10 \times 20 = 200$. The bandwidth would be 200 kHz and 100 kHz respectively. The second amplifier with a bandwidth of 100 kHz would dominate and the overall bandwidth would be no more than 100 kHz since, for example, a 150 kHz signal would pass through the amplifier with a bandwidth of 200 kHz but not through the amplifier with a bandwidth of 100 kHz.

Problems 5.3.1

(1) An OP-07 operational amplifier has a GBW of 0.5 MHz and a slew rate of $0.17\,\text{V}/\mu\text{s}$. If the closed loop gain is set to 15:
   (a) What is the expected bandwidth?
   (b) How many stages would be needed to produce a multistage module with a gain of 1000 and a bandwidth of 50 kHz? (*Hint*: find the gain of one stage with a bandwidth of 50 kHz and then work out how many of these stages would need to be cascaded.)
   (c) If the output signal amplitude is ± 13 V, what is the maximum frequency which can be passed without distortion?

(2) Two OP-07s are used to form a multistage amplifier as in Figure 5.3.6. Calculate the overall bandwidth of this circuit.

(3) What is the maximum frequency which an LM741-based buffer amplifier can handle without slew rate distortion if the output is not to exceed $5\,V_{RMS}$? (LM741 GBW $= 10^6$ Hz typically, slew rate $= 0.5\,\text{V}/\mu\text{s}$)

(4) A microphone amplifier needs to be built to boost signals from $0.8\,\text{mV}_{RMS}$ to $1\,V_{RMS}$ while maintaining a bandwidth of 20 kHz. How many 741-based stages would be needed?

Design a suitable circuit.

Figure 5.3.6

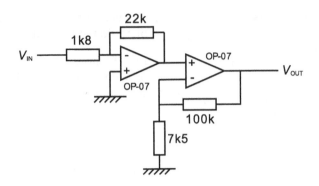

5.4 Variable gain amplifiers

In one respect, all amplifiers have a variable gain in that the gain is dependent on the value of the input and feedback resistors. This section looks into techniques for gain which is varied by an operator or user of the circuit.

Figure 5.4.1 shows a selection of amplifiers which offer variable gain. The problem with circuit (a) is that as the gain is adjusted, the bandwidth also alters because as we have seen, gain and bandwidth are not independent. If it is to be used as an a.c. amplifier, this could cause band limiting of the signal. The circuit is likely to be of limited use if the bandwidth changes every time the gain is altered!

Circuit (b) gets around this by introducing the gain control as a potentiometer leading into a fixed gain amplifier. This is preferable for a.c. circuits. If the shunting effect of $R1$ becomes noticeable, (causing a non-linear gain adjustment characteristic for the potentiometer) then the buffer amplifier could be introduced at XY.

A common control strategy for d.c. amplifiers which do not have to amplify particularly high frequencies is shown in (c). Here, there are two controls:

- Zero – to correct any zero offset from the voltage input;

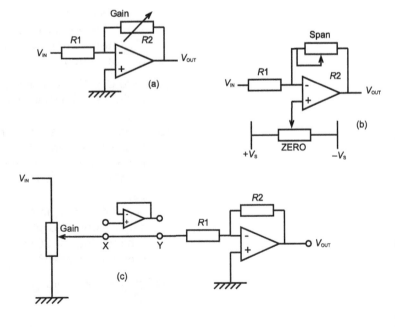

Figure 5.4.1 *Variable gain amplifiers*

Operational amplifiers (1)

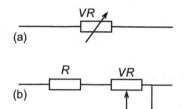

Figure 5.4.2 *Potentiometers*

- Span – sets the gain up to a full scale value, i.e. the span is the range of the output.

Potentiometers

Potentiometers have three connections – two at either end of the resistive track and one for the wiper. When a variable resistor VR is needed, it is recommended to connect the wiper to one of the track ends. The reason is that if a mechanical problem occurs with the wiper such as the wiper breaking, or dirt getting between the track and the wiper, then the resistor is never open circuited. It will fail to the maximum value of the variable resistor in a fault condition.

Figure 5.4.2 shows the general arrangement. R is a resistor which is sometimes included if it is important that the resistance never falls to zero. This could be used for a span control which sets the gain from a minimum to a maximum value.

5.5 Input offset errors

Input offset current

In an operational amplifier, when the input voltages are zero, a small bias current still flows into or out of the op-amp. These currents are the control currents for the input transistors of the op amp. Ideally, the current at the +IN input should be the same as the current at the −IN input. Manufacturing imperfections usually result in a very small but measurable difference. The magnitude of the **difference** of the two bias currents is called the input offset current:

$$I_{OS} = I_{B-} - I_{B+}$$

The input bias current is the average of the two bias currents. Both are specified at a particular temperature because with all semiconductor devices the parameters vary greatly with temperature.

When driven by high source impedances, the input bias and offset current effects swamp the effect of input offset voltage:

$$I_B = \frac{I_{B-} + I_{B+}}{2}$$

The input bias current creates a voltage drop across the input resistances of the op amp. The effect is to produce an unwanted offset voltage at the output.

To help reduce the effect of the bias currents, it is important to try to make the external resistance seen by both +IN and −IN signal inputs the same (Figure 5.5.1).

FET input op amps are markedly superior to bipolar op amps because of their higher input impedance. They have offset currents in the pA region as opposed to the nA experienced by the bipolar op amps. See Table 5.1.1 for some examples which show the differences

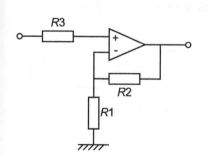

Figure 5.5.1 *Reducing offset error*

Input offset voltage

When both inputs of an op amp are connected together and to 0 V, it would be expected that the output would also be 0 V. This is rarely the case. V_{BE} mismatch in the input transistors causes a small

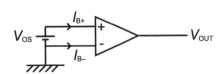

Figure 5.5.2 *Offsets*

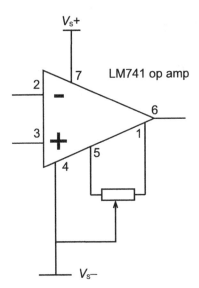

Figure 5.5.3 *Input offset voltage null circuit*

output to be present. The voltage which has to be applied to the input to cancel this out is called the input offset voltage. It also is specified at a particular temperature.

Figure 5.5.2 shows the small offset voltages and currents which have to be cancelled with external circuits.

With some operational amplifiers, the input voltage offset can be nulled with an external circuit such as is shown in Figure 5.5.3.

With amplifiers which do not have this facility, the approach used in Figure 5.4.1 must be used. The Zero control in this circuit can help null out the offset effect.

5.6 Common mode rejection ratio

When the same signal is applied to both inputs of an operational amplifer then, in the ideal case, no output would be produced. All practical amplifiers have a finite ability to reject these common mode signals. The figure of merit which indicates how well an amplifier can reject common mode signals is called the **common mode rejection ratio** (CMRR). It is defined as:

$$\text{CMRR} = 20\log_{10}\left[\frac{\text{Differential mode voltage gain}}{\text{Common mode voltage gain}}\right] \text{dB}$$

For example:

Amplifier	741	TL081	OP-77
CMRR (dB)	70	76	130

For the 741 amplifier, the ratio of differential:common mode gains is given by:

$$\text{antilog}_{10}\left(\frac{70}{20}\right) = 3162$$

If the typical differential mode gain is 1 000 000, then the common

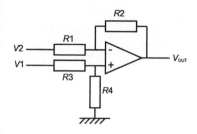

Figure 5.6.1 *The differential amplifier*

mode gain must be 316 before any feedback is applied (Figure 5.6.1).

In Section 5.2, the gain of this amplifier was shown to be:

$$V_{OUT} = \frac{R4}{R3+R4}\frac{R1+R2}{R1}V1 - \frac{R2}{R1}V2$$

If the signals are common mode, then $V1 = V2$, so:

$$A_{VC} = \frac{R4}{R3+R4}\frac{R1+R2}{R1} - \frac{R2}{R1}$$

If $V1 = V_D/2$ and $V2 = -V_D/2$, the differential mode voltage is V_D and the differential mode voltage gain becomes:

$$A_{VD} = \frac{1}{2}\left(\frac{R4}{R3+R4}\frac{R1+R2}{R1} + \frac{R2}{R1}\right)$$

$$\text{CMRR} = \frac{A_{VD}}{A_{VC}} = \frac{1}{2}\left[\frac{\frac{R4}{R3+R4}\frac{R1+R2}{R1} + \frac{R2}{R1}}{\frac{R4}{R3+R4}\frac{R1+R2}{R1} - \frac{R2}{R1}}\right]$$

$$= \frac{1}{2}\left[\frac{R4(R1+R2) + R2(R3+R4)}{R4(R1+R2) - R2(R3+R4)}\right]$$

$$= \frac{1}{2}\frac{R1R4 + R2R3 + 2R2R4}{R1R4 - R2R3}$$

In theory, with perfectly matched resistors, the CMRR is infinite. The best choice to make is:

$R1 = R3 = R$

$R2 = R4 = NR$ so that the differential mode gain is N

(as was shown in Section 5.2).

Perfect matching is difficult, and in practice the CMRR will not be all that high. Chapter 7 shows an improvement on this circuit in that it has a much better performance as a differential amplifier. This is to be the **instrumentation amplifier**.

6 Noise

Summary

Noise is an unwanted signal which appears in a system. It may be an impulse, a specific frequency, a set of harmonically related frequencies or range of frequencies. The causes can be external (in which case, the noise is usually called interference) or internal (from the electronic components themselves). This chapter will teach you to recognize the causes of noise, and suggest methods by which noise can be reduced.

6.1 Interference

Interference occurs when a signal is coupled onto the signal path through an accidental route. This could be:

- Directly via the power supply lines. (Full wave rectified mains supplies are a good source of 100/120 Hz signals, as discussed in Chapter 1.) Switch mode power supply (SMPS) frequency signals can also be conducted or radiated onto signal lines. SMPS operate typically from 20 kHz to 100 kHz and take advantage of the lower losses which occur at these higher frequencies. The switching impulses can generate frequency harmonics into the MHz region.
- Indirectly by **crosstalk**. This is a term which implies the coupling of unwanted signals onto the signal path of interest. Crosstalk pickup can be capacitive (electrostatic coupling) or inductive (electromagnetic coupling). Radio-frequency (r.f.) coupling could also be included in this category.

Interference can cause serious degradation to the signal information. However, with suitable correction techniques it usually can be reduced to negligible levels.

6.2 Reduction of power supply interference

The a.c. input should be protected from mains-borne transients. These could be in the form of voltage spikes or sudden changes of voltage which can be caused by heavy electrical loads switching on or off nearby. Typical of these might be from solenoid valves

Figure 6.2.1 *Mains input filter*

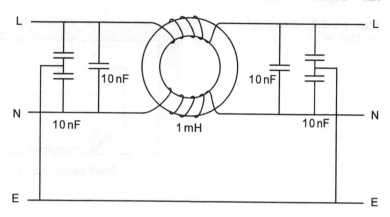

operating, an electric motor running with sparking contact brushes or fluorescent lights switching on. A mains input r.f. filter can be used to suppress these. This device has components inside which are typical of those shown in Figure 6.2.1.

The r.f. filter should be mounted where the supply cord enters the body of the equipment to prevent the sharp transients coupling within the case to other parts of the circuitry. Some a.c. power connectors have filters built in as part of the connector/switch/fuse assembly. The toroidal ring inductors are wound so that the magnetic fields generated by high frequencies in the supply lines cancel in the ferrite ring. The capacitors present a low impedance to high freqencies and shunt these signals to ground. Sometimes transient suppressors are used as well. These are small, inexpensive electronic components which present a high impedance below a certain voltage, and a low impedance above. They are rated by voltage (available from 5.5 V to 500 V) and either by peak (pulsed) current capacity or by their energy dissipation properties. They are used to protect the equipment from mains-borne transients.

Both of these protection methods are bidirectional. So, if the equipment turns out to be a good noise generator itself, the supply system and adjacent equipment is protected against noise leaking out of the noisy circuit.

Once the mains a.c. is inside the enclosure, the cable should be routed directly to the transformer, avoiding any of the signal processing circuitry; especially the input wiring. The transformer could be of the shell type construction or, preferably, the more expensive toroidal construction as shown in Figure 6.2.2.

The magnetic path of the toroid is constrained inside the primary and secondary windings and there is much less chance of magnetic leakage which could couple to any nearby low voltage circuitry.

Whichever transformer is used, its connecting leads should be twisted together as in Figure 6.2.3. Each twist is a small loop which will magnetically have voltages induced in it. However, each successive loop has the opposing polarity of voltage induced, so the overall signal picked up cancels with the preceding loop.

Decoupling capacitors

The rectified low voltage output of the power supply unit is decoupled by capacitors to remove high frequency signals super-

Figure 6.2.2 *A toroidal transformer*

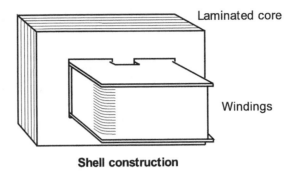

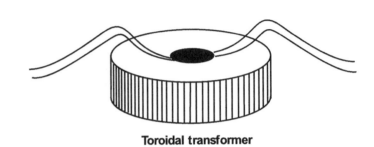

Figure 6.23 *A twisted cable arrangement*

imposed onto the d.c. In digital design, because of the effect of fast switching speeds, the V_{CC} power supply can be significantly disrupted. It has long been a maxim that each digital integrated circuit should be protected with a closely coupled 0.1 µF capacitor. If the mains transformer has to be mounted close to sensitive electronics, then it may be necessary to screen the transformer (or electronics) with a high-mu (high magnetic permeability) metal screen.

6.3 Ground loops

In terms of the power distribution, there is another problem which can occur with the 0 V line. Ideally, this is a wire (or a track on a printed circuit board) which is the 0 V reference for the whole circuit. However, in some designs, where there is appreciable current flowing (or the 0 V line has too high a resistance), then $V = IR$ comes into play. The circuit of Figure 6.3.1 shows the general idea.

When the ground line returns current from the power output stage, a voltage develops across the apparent resistance r. This is going to upset the virtual earth of the amplifiers somewhat. One way around this problem is the 'star earth' connection as shown in Figure 6.3.1(b). Incidentally, this is one of the few occasions where the convention is flouted involving orthogonal connecting lines on a circuit diagram. The angled lines are introduced to emphasize the common ground point.

Figure 6.3.1 *Ground loops*

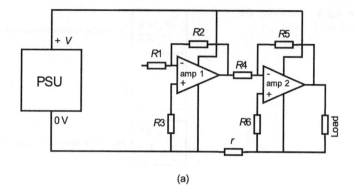

(a)

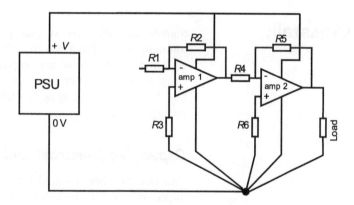

Voltage sense circuits

To get around the problem of voltage drop in supply cables, power supply units which are expected to deliver appreciable current are often fitted with a set of sense terminals as in Figure 6.3.2.

The heavier V_{out} supply lines are routed in the normal way to the electronic unit. The separate sense wires have a high input impedance back in the voltage regulator stage of the power supply unit – and because very little current flows in them, they do not suffer a significant voltage drop between end A and end B. This helps maintain a constant voltage at the PSU as it compensates for any voltage drop in the V_{out} supply lines.

If there are a number of power supply units, the 0 V return lines should not be commoned together at each power supply unit. Each supply pair should be run as a twisted pair to the equipment supplied, and the 0 V lines commoned there as shown in Figure 6.3.3.

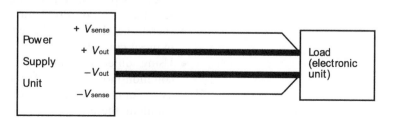

Figure 6.3.2 *Power supply unit with voltage sense circuit*

Figure 6.3.3 *Handling 0V common lines*

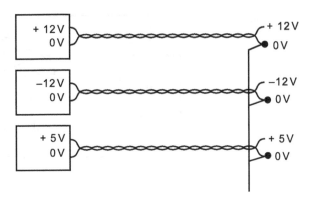

6.4 Crosstalk

Signals can end up at one point in a circuit after coupling unintentionally from another. This is called crosstalk. The two coupling mechanisms are capacitive and inductive. Capacitive coupling occurs when there is a high dV/dt change in a signal line. Inductive coupling occurs when there is a high dI/dt change in the signal line.

Capacitive (electrostatic) coupling

This occurs when a signal line runs close to another line of high impedance. When the voltage changes rapidly in a transmitter line, the victim line follows to some extent. The cure involves:

- Moving the wires apart, if possible. This reduces the capacitance between the wires.
- Reducing the impedance of the victim circuit. The impedance determines the amount of pickup between the wires.
- Introducing a screen around one or both of the wires. The screen could be a metal enclosure or a mesh as is used in coaxial cables. Not quite as effective, but of some worth, is the technique of moving both emitter and victim wires close to a ground plane. This is why, on some printed circuit boards, you can see large areas of conductor.

Magnetic (inductive) coupling

The magnetic field is generated when the current changes rapidly in the transmitting wire. The victim circuit will pick this up quite easily. The cure involves:

- Separating the circuits. The magnetic coupling will be reduced.
- Reducing the loop area of the victim circuit. The induced signal is largely dependent on the area of the coil since it acts like the secondary coil of a transformer.
- Increasing the circuit impedance. This reduces the amount of induced current.
- Using differential amplifiers in the victim circuit to cancel common-mode effects. Amplifiers will pick up the interference as a common mode signal. Using differential techniques will eliminate these.

- Using magnetically shielded wires or screens. These screens must be made from special mu-metal sheets which have a high permeability to magnetic fields.

In general, a magnetic screen will also act as an electrostatic screen, but **not** vice versa!

Coaxial cable screens

Care must be taken when dealing with any coaxial screen, i.e. where and how to ground it. The signal is usually the core wire of the cable, with the screen providing some protection around it. The screen can be grounded as shown in Figure 6.4.1.

However, if the ground points have a small resistance between them, then earth currents can flow, and a small voltage will be introduced. The intrusion of this signal depends on the amplitude of the wanted signal between the equipment. One solution to this is to use a screened twisted pair cable to interconnect the two pieces of equipment. The second wire of the pair is connected to the screen at the transmitter end as shown in Figure 6.4.2.

The differential amplifier at the receiver then amplifies the wanted signal and rejects the common mode signal. This form of the differential amplifier would usually be inadequate to reject all of the common mode noise, and it would be better to see a dedicated instrumentation amplifier (see Chapter 7).

In really bad cases of noise pickup via the connecting lead, perhaps made worse by earth loop currents, then there are devices called **isolation amplifiers**. These are amplifiers which require two sets of power supplies – one for the input (transmitter) stage and one for the output (receiver) stage. They exhibit complete electrical isolation from the input to the output. Figure 6.4.3 shows the schematic layout. Transmission from the input to the output is either by light (LED and photodiode) or magnetism (coupling transformer).

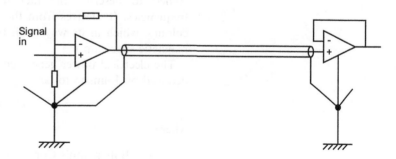

Figure 6.4.1 *A simple screen grounding*

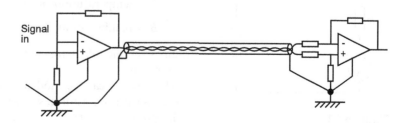

Figure 6.4.2 *A screened twisted pair*

Figure 6.4.3 *An isolation amplifier*

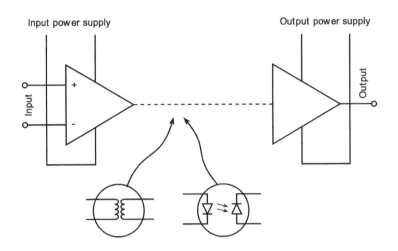

6.5 Internal noise

Internal noise exists in any electrical conductor or component. It is caused by the movement of discrete charge carriers either drifting randomly, or moving under an applied external electrical field, or crossing a semiconductor pn junction.

Johnson noise

A simple resistor consists of electrons moving randomly about, and at any one instant there is a finite probability that there will be more electrons at one end than the other. When this random movement causes a temporary bunching up of the charge carriers, there will be a small, but measurable, voltage across the resistor. This is called the thermal, or Johnson, noise. As the temperature increases, so does the thermal agitation of the electrons – the noise voltage increases. Only at the absolute zero temperature of 0 Kelvin (or $-273.15\,°C$) will the noise voltage be reduced to zero. Also, since it is random motion, all frequencies are equally likely, so the wider the bandwidth of the measuring equipment, the more noise is measured and the larger the noise voltage. The term **white noise** has been coined to describe the fact that Johnson noise contains all frequencies. (It derives from the fact that white light contains all colours, which in its way is all frequencies of the optical electromagnetic spectrum.)

The electrical power generated by a conductor (or resistor) was deduced by Johnson to be:

$$P = kTB$$

where

k = Boltzmann's Constant (1.38×10^{-23} Joules/Kelvin)

T = temperature in Kelvin

B = bandwidth in Hertz

Many scientific calculators have Boltzmann's constant programmed into memory.

The resistor can be modelled as in Figure 6.5.1 which shows a noiseless (ideal) resistor in series with a noise voltage generator. The noise power is independent of the value of the resistor. We need to

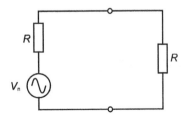

Figure 6.5.1

determine how much of that power can be coupled to an external circuit. Recall that the maximum power transfer theorem states that maximum power can be coupled from a source to a load when the impedance of the load matches the impedance of the source.

In Figure 6.5.1, the current going around the circuit would be:

$$I = \frac{V_N}{R+R} \quad \text{or} \quad \frac{V_N}{2R} A$$

The power is given by:

$$P = I^2 R = \frac{V_N^2}{4R^2} R \quad \text{which is} \quad \frac{V_N^2}{4R} W$$

$$P = kTB = \frac{V_N^2}{4R}$$

Thus the noise voltage generator is given by:

$$V_N = \sqrt{4kTBR} \text{ V}$$

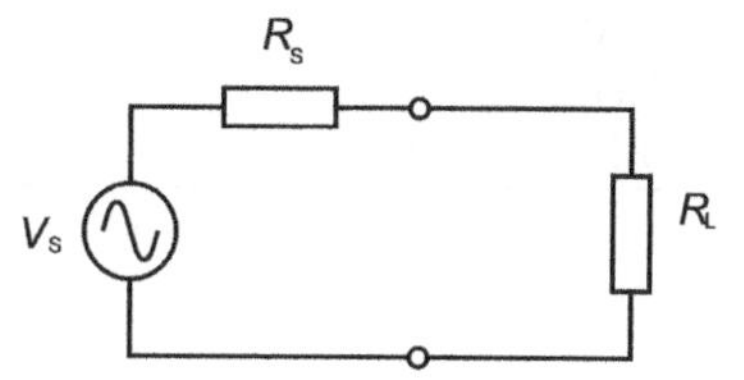

Figure 6.5.2 *Load matching*

Mathematics in action

A generator V_S with internal resistance R_S is coupled to a load of R_L. The power dissipated in the load is:

$$P_L = \left(\frac{V_S}{R_S + R_L}\right)^2 R_L$$

This is maximum when:

$$\frac{dP_L}{dR_L} = 0 \quad \frac{dP_L}{dR_L} = \frac{(R_S + R_L)^2 - R_L 2(R_S + R_L)}{(R_S + R_L)^4}$$

This is zero when the numerator is zero:

$$(R_S + R_L)^2 = 2R_L(R_S + R_L)$$

$$R_S + R_L = 2R_L$$

$$R_S = R_L$$

i.e. maximum power is transferred when:

Source resistance = load resistance

The noise voltage generated is a random process, so the amplitude given by the equation is also bound by probability relationships. Analysis and practical observation has shown that the noise amplitude distribution has a Gaussian characteristic as shown in Figure 6.5.3. This is a graph which expresses the probability of a particular noise voltage being generated at any instant.

Given the formula:

$$V_N = \sqrt{4kTBR} \text{ V}$$

it is possible to work out the average voltage generated at any temperature. For example, the noise voltage generated by a 1k

Figure 6.5.3 *A Gaussian distribution*

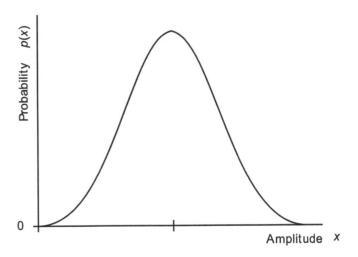

resistor operating into a bandwidth of 20 kHz at room temperature would produce a noise voltage of 0.57 µV. As the resistor value alters, so will the noise voltage, in accordance with the following table.

Resistance	Noise voltage at room temperature
100	0.18 µV
1k	0.57 µV
10k	1.8 µV
100k	5.7 µV
1 M	18 µV

To reduce Johnson noise:

- work with low resistances;
- reduce the operating bandwidth to the minimum necessary;
- lower the temperature.

Problems 6.5.1

(1) Calculate the Johnson noise generated by a 15k resistor when operating at room temperature (294 K) into a bandwidth of:
 (a) 20 kHz;
 (b) 100 kHz.
(2) (a) Calculate the Johnson noise generated by a 100k resistor operating into a 50 kHz bandwidth at room temperature of 290 K.
 (b) What temperature drop is required to lower the noise by a factor of 4?
(3) What is the Johnson noise generated by a 50 Ω resistor at 100 °C into a bandwidth of 150 kHz?

Shot noise

The previous, Johnson, noise occurred whether or not the resistor was connected into a circuit. Another physical mechanism occurs when current flows through that resistor. So-called **shot noise** arises because current flow is the movement of discrete packages of electrical charge. As they travel through a conductor or resistor, there are random fluctuations of their movements. The fluctuating noise current is given by:

$$I_N = \sqrt{2qI_{DC}B}$$

where

I_N is the current shot noise

q is electron charge (1.6×10^{-19} coulombs)

I_{DC} is the d.c. bias current

B is the bandwidth

Measurement of the shot noise due to a 1 A current in a 20 kHz bandwidth would give 80 nA contribution to the overall noise. In characteristic, the noise is white (broadband) and Gaussian.

Shot noise is also evident when charge carriers cross a pn junction in a diode. The random timing of the crossings of the charge carriers is related to the current flow and the bandwidth of the measuring equipment as given in the equation:

$$I_N = \sqrt{2qI_{DC}B}$$

Shot noise can be measured in bipolar junction transistors because of the base–emitter/base–collector junctions. Ideal FETs, being unipolar and with no junctions crossed by charge carriers, are almost free of shot noise (in practice, there is a small amount of shot noise generated by reverse conduction from the channel into the gate).

Problems 6.5.2

(1) Calculate the shot noise due to a current of 1 mA flowing through a diode when measured by an instrument with a 10 kHz bandwidth. (This noise can be reduced by operating a transistor with low bias currents. The smaller the I_{DC}, the smaller the noise.)

(2) Calculate the shot noise when the bias current of the previous problem is reduced to 100 nA.

Flicker noise or 1/f noise

The alternative name of $1/f$ noise comes about because the noise generated in this mechanism is not white. It is called **pink** noise, and is higher at low frequencies. It occurs in resistors because of the different materials from which resistors are made and from the different resistor construction methods (i.e. how the wires are bonded to the resistance material).

For example:

Wire wound	0.01–0.2 µV
Metal film	0.04–0.2 µV
Carbon film	0.05–0.3 µV
Carbon composition	0.2–3.0 µV

Burst (popcorn) noise

This is another type of low frequency noise which is known to exist, but which is related to the purity of the semiconductor manufacturing process. Measurements on a device prone to popcorn noise would show a sudden, short-lived shift in the bias current level which quickly returned to its original state. The popping burst of noise gave rise to the term popcorn noise. It increases with bias current and is proportional to the inverse of the square of the frequency:

$$\text{Burst noise} \propto \frac{1}{f^2}$$

6.6 Noise models

A very simple circuit such as that shown in Figure 6.6.1(a) could be represented by its model in Figure 6.6.1(b). In this way, the circuit noise could be evaluated.

The diode and the resistor will both generate noise.

In the resistor, Johnson noise will be generated to a value of

$$V_N = \sqrt{4 \times 1.38 \times 10^{-23} \times 10 \times 10^3 \times 298 \times 10 \times 10^3} = 1.3\,\mu V$$

The diode has a voltage drop of approximately 0.6 V, so its forward bias current would be:

$$I_F = \frac{10 - 0.6}{10\,000} = 940\,\mu A$$

The dynamic forward resistance is given approximately by:

$$R_F \approx \frac{0.026}{I_{DC}}$$
$$= 28\,\Omega$$

With such a small dynamic resistance as this, the Johnson noise will be fairly small. The bigger contribution will be from shot noise:

$$I_N = \sqrt{2 \times 1.6 \times 10^{-19} \times 940 \times 10^{-6} \times 10\,000} = 1.7\,nA$$

Given any transistor or operational amplifier circuit, it is possible to estimate the expected noise at any temperature and bias level. However, with any but the most simple circuits, this would normally be done with a software modelling tool which can work quickly with all the noise signal generators which appear on behalf of any circuit.

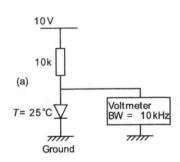

Figure 6.6.1 *A simple noise model*

6.7 Signal to noise ratio

The signal to noise ratio is a figure of merit for systems which is measured in dB and defines, as the name implies, a figure for the ratio of signal power to noise power. An amplifier will equally amplify any signal or noise which is presented at its input. However, because of the electronic components, noise in introduced inside the amplifier so the signal to noise ratio at the output is always smaller than the signal to noise ratio at the input:

$$\frac{S_{out}}{N_{out}} < \frac{S_{in}}{N_{in}}$$

Noise figure (noise factor)

This is another term you may hear when dealing with electronic noise. The noise figure is defined as the ratio of the input signal to noise ratio to the output signal to noise ratio:

$$N_F = \frac{(S/N)_{in}}{(S/N)_{out}}$$

This is normally expressed in dB.

Transistors have a noise performance which looks typically like the circuit of Figure 6.7.1.

These somewhat complex diagrams show the lines of constant noise figure (in this case for bipolar junction transistor type BC549). It relates the noise figure as a function of the collector current and source resistance. You can see that to minimize the noise figure, a compromise has to be drawn between collector current bias and input resistance.

The noise model for a transistor is often conveniently worked out with the circuit of Figure 6.7.2.

The voltage and current noise generators (for a BCY71) have outputs which are described in graphs shown by Figure 6.7.3.

The voltage noise is produced by two mechanisms:

- the Johnson noise in the base spreading resistance;
- shot noise generated by the collector current which subsequently generates a voltage across the emitter resistance.

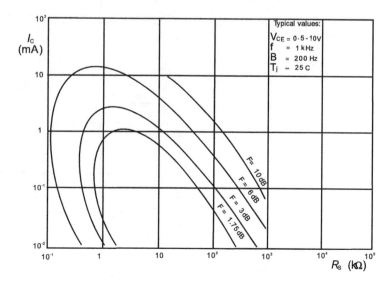

Figure 6.7.1 *Transistor noise performance*

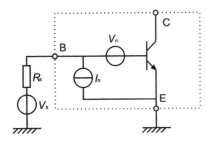

Figure 6.7.2 *Transistor noise model*

The noise voltage V_N drops with increasing I_C. The current noise is produced by shot noise caused by current flowing in the base. I_N rises with increasing I_C.

Noise temperature

Noise temperature is a term which is sometimes used instead of noise figure. The concept is that a noisy circuit generating output noise power PN could be considered to be a perfect amplifier with an output resistor generating noise as if at a temperature of T_N:

$$T_N = \frac{P_N}{A_P k B}$$

where

$$k = \text{Boltzmann's constant}$$
$$A_P = \text{power gain}$$
$$B = \text{bandwidth}$$

The noise temperature of the output resistance is not necessarily at the actual temperature. It could be smaller (for a low noise amplifier) or larger (for a noisy amplifier).

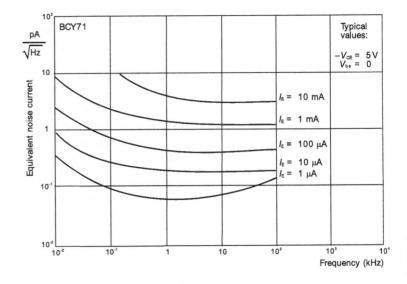

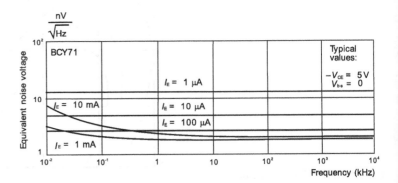

Figure 6.7.3 *Noise generators*

7 Operational amplifiers (2)

Summary

This chapter looks at applications of operational amplifiers in the real world. It also looks at more specialized devices which could be considered under the umbrella term operational amplifier. By the end of the chapter, you should be able to recognize the advanced circuit configurations, and be able to design these applications with confidence.

7.1 The instrumentation amplifier

In Chapter 5, we looked at a differential amplifier. While suitable for very simple applications, its performance can be improved. The instrumentation amplifier is a configuration using three operational amplifiers as shown in Figure 7.1.1. It has the following features:

(i) high input impedance;
(ii) high CMRR;
(iii) precision high gain.

For convenience, it is sometimes given the symbol of Figure 7.1.2.
The gain of Figure 7.1.1 is given by:

$$V_{\text{OUT}} = (V1 - V2)\left(\frac{R1 + 2R2}{R1} \times \frac{R4}{R3}\right)$$

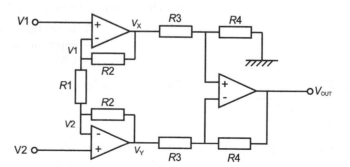

Figure 7.1.1 *The instrumentation amplifier*

Instrument amplifier analysis

Assume that the operational amplifiers have a high input impedance, so that negligible current enters the input terminals. The V− input is at the same potential as the V+ input.

The current flowing from V_X to V_Y is given by:

$$I = \frac{V_X - V_Y}{R2 + R1 + R2} = \frac{V_X - V_Y}{R1 + 2R2}$$

The current flowing through R1 is also related to (V1 − V2) as in:

$$I = \frac{V1 - V2}{R1}$$

Equating these:

$$\frac{V1 - V2}{R1} = \frac{V_X - V_Y}{R1 + 2R2}$$

$(V_X - V_Y)$ are the inputs to a differential amplifier as described in Chapter 5.

$$V_{OUT} = (V_Y - V_X)\frac{R4}{R3}$$

The overall gain of the instrumentation amplifier is:

$$V_{OUT} = (V1 - V2)\left(\frac{R1 + 2R2}{R1} \times \frac{R4}{R3}\right)$$

Figure 7.1.3 shows a method of adjusting the gain. If the op amps are used from a monolithic package (all ICs on one piece of silicon, in one package) then they will track temperature changes. That is, when the temperature changes, then all operational amplifiers are equally affected. This helps minimize the sensitivity of the instrumentation amplifier to changes in temperature. Figure 7.1.4 shows a popular pinout of one such package. It would apply to such op amps as the LM324, TL064, TL074, TL084, etc.

The performance can be improved even further by using the more precision matching obtained with a custom instrumentation amplifier such as the Burr Brown INA116 of Figure 7.1.5. This instrumentation amplifier uses matched operational amplifiers and matched resistors to give a performance of:

Input bias current	3 fA
Input offset voltage	2 mV
CMRR	84 dB

The gain is set by one external resistor to give a gain of:

$$A_V = 1 + \frac{50\,k\Omega}{R_G}$$

The guard inputs are part of a technique common to precision instrumentation devices. The guard inputs help achieve the low

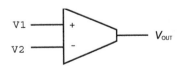

Figure 7.1.2 *An instrumentation amplifier*

Figure 7.1.3 *Varying the gain of an instrumentation amplifier*

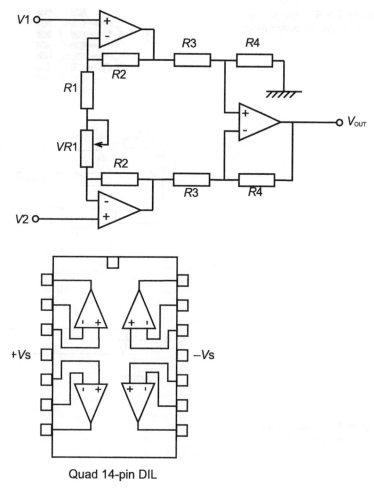

Figure 7.1.4 *A quad op amp package*

Quad 14-pin DIL

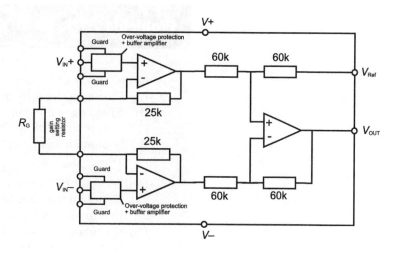

Figure 7.1.5 *The Burr Brown INA116*

input bias and would be connected on a printed circuit board as shown in Figure 7.1.6(a). Physically, the guard pins are adjacent to the input pins which they are guarding on the integrated circuit. If guarding back to the signal source is required, the guard could be extended via the coaxial cables as shown in Figure 7.1.6(b).

An alternative approach is to use the guard pins to drive a buffer amplifier the output of which would drive the screen of the coaxial cable as in Figure 7.1.7. This is effective with fast input signals.

Figure 7.1.6 *Use of the instrumentation guard pins*

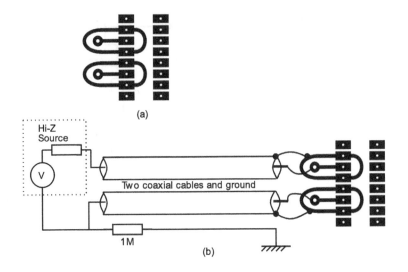

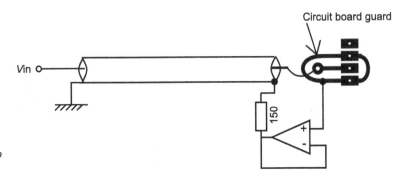

Figure 7.1.7 *Using guard pins to drive the input coaxial cable*

Problems 7.1.1

(1) In the circuit of Figure 7.1.8:
   (a) Calculate the gain of the amplifier.
   (b) Show how the gain could be continuously varied over a range of 10–25.
   (c) Design an instrumentation amplifier with a gain of 100.

(2) In the circuit of Figure 7.1.9:
   (a) Calculate the gain of the amplifier when *VR* is set to a minimum value.
   (b) Calculate the gain of the amplifier when *VR* is set to a maximum value.

Bridge measurement methods

Instrumentation amplifiers are often used with measurement bridges. These are versions of Wheatstone Bridges where one or more of the bridge elements have resistors which alter according to

Figure 7.1.8

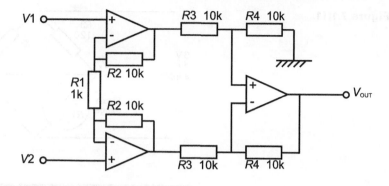

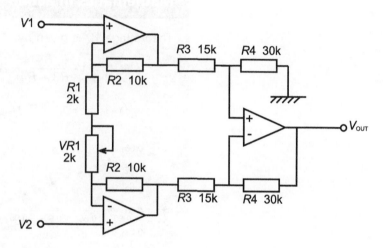

Figure 7.1.9

changes in some physical parameter. For example, the resistance could be altered by:

Temperature	Platinum resistance thermometer (or thermistor)
Mechanical strain	Strain gauge
Light	Photoconductive cell
Magnetism	Magneto-resistor

Figure 7.1.10 shows a Wheatstone Bridge where resistor R4 has its resistance altered according to whatever parameter it has been optimized to measure.

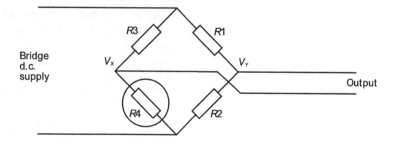

Figure 7.1.10

Figure 7.1.11

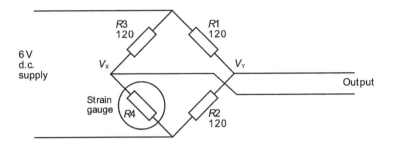

Mathematics in action

The voltage V_X is given by:

$$V_X = V_S \frac{R4}{R3 + R4} \quad V_Y = V_S \frac{R2}{R1 + R2}$$

$$V_{Y-X} = V_Y - V_X = V_S \frac{R2}{R1 + R2} - V_S \frac{R4}{R3 + R4}$$

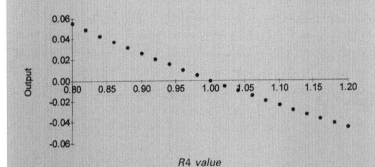

The graph shown was derived from the situation where $R1 = R2 = R3$ and $R4$ varied from 80% to 120% of the value of $R1$. The output is not linear when $R4$ varies over such a large percentage. Care must be taken when designing a Wheatstone measurement bridge that the value of $R4$ only varies over a small percentage of the other bridge resistors. Then good accuracy can be achieved.

Example 7.1.1

A 120 Ω strain gauge is to measure the strain in a cantilever beam. It is included as part of a 6 V d.c. Wheatstone Bridge with three precision 120 Ω resistors and the one strain gauge. Under maximum strain, the gauge resistance alters to 120.6 Ω. Design an instrumentation amplifier which will output 0–1 V from minimum to maximum strain.

The output at V_Y is given by:

$$V_Y = 6\,V \times \frac{120}{120 + 120} = 3\,V$$

while the voltage at V_X is given by:

$$V_X = 6\,\text{V} \times \frac{120.6}{120 + 120.6} = 3.0075\,\text{V}$$

The differential output will rise from 0 V to 7.5 mV, and this needs to be amplified to produce an output from 0 V to 1 V. Hence the gain must be:

$$A_V = \frac{1}{0.0075} = 133$$

If we use the circuit for the instrumentation amplifier of Figure 7.1.1, the gain would be:

$$A_V = (V_1 - V_2)\left(\frac{R_1 + 2R_2}{R_1}\right)\frac{R_4}{R_3}$$

Let $R_4 = R_3 = 10\text{k}$ and $R_1 = 1\text{k}$. Then:

$$A_V = 133 = (0.0075)\left(\frac{1 + 2R_2}{1}\right)\frac{10}{10}$$

$$\frac{133}{0.0075} = 1 + 2R_2$$

$$R2 = 8.86\,\text{k}\Omega$$

Problems 7.1.2

(1) A 120 Ω strain gauge is bonded to a load cell to measure the strain in the cell and hence the forces to which it is subjected. The rest of the Wheatstone Bridge is constructed from precision 120 Ω resistors and connected to a 2 V bridge supply. Design an instrumentation amplifier which will provide a 0–1.5 V output signal when the gauge has its resistance altered from 120 Ω to 119.2 Ω at full load.

(2) In question 1, if the unstressed resistance is actually 120.1 Ω, show how the offset could be nulled within the instrumentation amplifier.

(3) A PT100 temperature sensor is part of a temperature measurement system. The sensor has a resistance of 100 Ω at 0°C and has three other precision 100 Ω resistors to make the Wheatstone Bridge (which is powered from a 5 V supply). The temperature is to be displayed on a four-digit display that requires 0.1 mV for the least significant digit. (That is, it would require 199.9 mV for a full scale display. More correctly, a display which can read from 0000 to 1999 is called a $3\frac{1}{2}$ digit display.) Design an instrumentation amplifier which will display the correct temperature (between 0 and 70°C) when the PT100 has a temperature coefficient of 0.385 Ω per °C.

(4) A thermistor is to be used as the basis of a fish tank temperature display unit. The thermistor has a linear change of resistance with temperature over a small range, and is known to have a resistance of approximately 4200 Ω at 18 °C and 4000 Ω at 25 °C. The display is an analogue moving coil meter with coil resistance of 3 kΩ and sensitivity of 100 µA for a full scale deflection (FSD). Design an instrumentation amplifier that will interface between the Wheatstone Bridge and the meter. The meter is to indicate 18 °C minimum and 25 °C maximum.

7.2 Chopper stabilized amplifiers

In Chapter 5 we covered the limitations caused by input offset current and voltage. The problems become particularly difficult when dealing with very small signals which vary slowly with time; for example, signals from transducers such as strain gauges or thermocouples. Offset effects can be nulled out, but a problem still exists due to the drift of V_{OS} and I_{OS} with temperature and time. The chopper stabilized amplifier is a device which uses techniques to reduce these effects when amplifying low frequency, low amplitude signals. The technique is not new. It was developed during the valve era and used a mechanical switch to chop the slowly varying d.c. into an a.c. signal. The resultant a.c. was amplified using conventional techniques and then rectified back to d.c. The limitations of the technology of the time put severe restrictions on the bandwidth which could be obtained. The chopping frequency had to be bigger than the largest frequency of interest and since the chopping was mechanical (50–200 Hz) the technique was limited to slowly varying signals. It was still an effective technique and was extensively used for thermocouple amplifiers. Figure 7.2.1 shows the method.

The oscillator drives the switch which converts the input into an a.c. waveform. This schematic omits some of the sophistication needed to reconstruct the output without significant noise. The oscillator shown would be a transistor switch these days, but might have been a coil or even motor driven reed relay in the days of valve amplifiers.

The phrase chopper stabilized amplifier refers to a different technique nowadays. The electronics are not quite so prone to drift, so the chopping technique is used to minimize another limitation of operational amplifiers: offset. An alternative name is

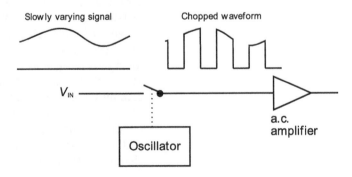

Figure 7.2.1 *Chopper stabilized amplifier*

Figure 7.2.2 *Commutating auto zero amplifier*

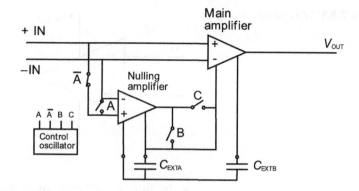

the **commutating auto zero** amplifier, sometimes abbreviated to CAZ. There is a spectacular improvement in performance, but the most noticeable change is that the amplifier has a bandwidth which is not affected by the chopping frequency. The chopper stabilized amplifier actually contains two identical amplifiers which, since they are fabricated on the same piece of silicon at the same time, will have very similar characteristics. The main amplifier is connected between the input and output pins of the device. The other amplifier is a nulling amplifier as shown in Figure 7.2.2.

The improvement can result in devices with typical parameters as shown in Table 7.2.1.

Table 7.2.1

Device	ICL7650	MAX420
VOS (μV)	5	5
VOS temperature drift (μV/°C)	0.1	0.05
IB (pA)	20	100
AVOL	120	120
CMRR (dB)	110	120
Slew rate (V/μs)	2.5	0.5

One of the industry standards which has evolved is the ICL7650 and Figure 7.2.2 shows this in schematic form. The control oscillator free runs at approximately 200 Hz (although it can be set to other frequencies) The main amplifier does its normal job of amplifying the difference between the +IN and −IN signals. **Nulling** is the process of applying a small voltage to the amplifier to counteract the offsets. The nulling amplifier alternately nulls itself and then the main amplifier at a frequency determined by the oscillator. The offset voltages are stored on the two external capacitors C_{EXT} which are recommended as 0.1 μF. The nulling connections are high impedance MOSFET gates which do not take much current from these external capacitors. The drawback of many types of these devices are that:

- there may still be a detectable switching noise on the output;
- they do not have a particularly good output current drive capability;
- inputs above the supply voltage may damage them;

Figure 7.3.1 *The comparator*

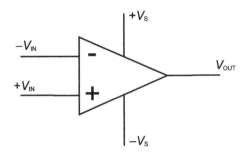

- the supply voltage is lower than with conventional operational amplifiers.

7.3 Comparators

A comparator is a very simple form of analogue to digital converter. There are two analogue inputs, and the output changes state from 0 to $+V$ or from $-V$ to $+V$ when the $+$IN input is greater than the $-$IN input. The symbol used is the same as that of an operational amplifier – indeed, it is possible to use an op amp as a comparator.

The disadvantages of using an op amp as a comparator are:

(1) The output will only switch between the output saturation voltages of $-V_{SAT}$ and $+V_{SAT}$. Usually, these fall 1 to 2 V short of the supply voltages $-V_S$ and $+V_S$.
(2) The inputs are susceptible to noise. A slightly noisy signal on one input can cause the output to switch states.

Like most comparators, the output of the LM311 is open collector as shown schematically in Figure 7.3.2. It must have a pull-up resistor or some other form of load from the output to a positive supply.

Pin 1 defines the lower output voltage to which the emitter of the transistor is connected. When $+$IN is greater than $-$IN, the transistor switches ON, and so the output will go LOW. The advantage of this arrangement is that the output can be connected to voltages other than the supply voltage (up to 50 V for the LM311). This would allow the use of lower or higher voltage outputs for circuitry such as level shifters, display drivers, relay drivers, etc. The output of a comparator is capable of sinking far more current than operational amplifiers. (An output **sink** is when

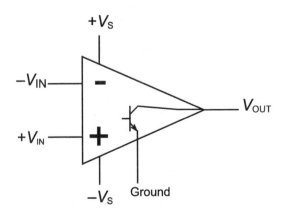

Figure 7.3.2 *Comparator output stage*

Figure 7.3.3 *Switching transients*

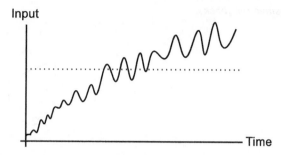

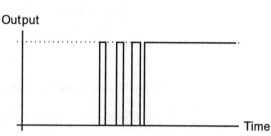

current goes **into** the output. An output **source** is when current comes **out** from the output. An LM311 can sink 50 mA.)

Table 7.3.1 shows the properties of a small range of common comparators.

Table 7.3.1 *Typical comparators*

Type	Supply max. (V)	Supply min. (V)	Switching speed (ns)	V_{OUT} max. (V)	I_{SINK} max. (mA)
LM311	30	5	200	50	50
NE521	10	9	20	V_S	115
MAX931	11	2.5	12	V_S	40
AD9696	10	5	4.5	V_S	20
CA3290	36	5	1000	36	30

A comparator only requires a few mV between the input signals before switching can take place. This allows the possibility of chatter or switching transitions (as in Figure 7.3.3).

A noisy signal will cross the threshold voltage and cause the transients shown. At each crossing, the output changes state. To eliminate this, hysteresis is introduced by positive feedback.

In the circuit of Figure 7.3.4, the threshold voltage at V_X will be different depending whether the output is HIGH or LOW. To show this, when V_{OUT} is HIGH (V_{OUTH}):

$$\frac{V_{OUTH} - V_{REF}}{R1 + R2} = \frac{V_X - V_{REF}}{R1}$$

$$V_X = V_{REF} + (V_{OUTH} - V_{REF})\frac{R1}{R1 + R2}$$

When V_{OUT} is LOW (V_{OUTL}):

Figure 7.3.4 *Positive feedback*

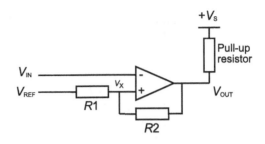

$$\frac{V_{REF} - V_{OUTL}}{R1 + R2} = \frac{V_{REF} - V_X}{R1}$$

$$V_X = V_{REF} - (V_{REF} - V_{OUTL})\frac{R1}{R1 + R2}$$

The hysteresis is the difference in these two values:

$$\text{Hysteresis} = V_{REF} + (V_{OUTH} - V_{REF})\frac{R1}{R1 + R2} - V_{REF}$$

$$+ (V_{REF} - V_{OUTL})\frac{R1}{R1 + R2}$$

$$= (V_{OUTH} - V_{OUTL})\frac{R1}{R1 + R2}$$

In the circuit of Figure 7.3.4:

$$V_{OUTL} = 0\,V \qquad R1 = 10k$$
$$V_{OUTH} = 10\,V \qquad R2 = 22k$$
$$V_{REF} = 5\,V$$

The one threshold level works out to 6.56 V and the other to 3.44 V. By subtraction or formula, the hysteresis works out to 3.12 V. Figure 7.3.5 illustrates the effect that this will have.

The comparator will switch when the $V_{IN}+$ input voltage is greater or smaller than the $V_{IN}-$ input voltage. When $V_{IN}+ > V_{IN}-$, the output transistor switches ON and the output will go LOW. When $V_{IN}+ < V_{IN}-$, the output transistor switches

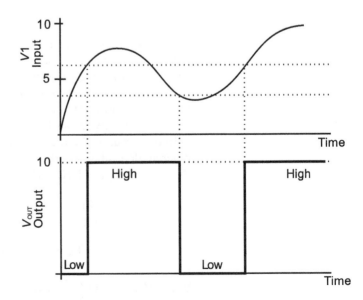

Figure 7.3.5 *A comparator with positive feedback*

OFF and the output is pulled high by the external pull-up resistor. The effect of hysteresis is to separate the switching point as shown in Figure 7.3.5.

If a comparator is used with the hysteresis resistors, then the signal source must be capable of providing the current through $R1$ without compromising the measurement accuracy. Sometimes a unity gain buffer amplifier might be required on the $V_{IN}+$ input.

> ### Example 7.3.1
>
> The circuit of Figure 7.3.6 shows a low-light alarm. Design a hysteresis comparator so that the LED switches OFF when V_D reaches 6.5 V and ON when V_X reaches 5.5 V.
>
> To make the problem easier, the bridge components $R3$ and $R4$ have already been provided, so there only remains $R5$, $R6$ and $R7$ to calculate. ($R1$ and $R2$, it is assumed, are chosen to produce the necessary output voltage at the appropriate light levels. A photoresistor or more properly a photoconductive cell has a property whereby the resistance falls as light increases. Hence if the light level falls, the resistance increases and so will the voltage drop across it.)
>
> $$6.5 = 6.0 + (12 - 6.0)\frac{R1}{R1+R2}$$
>
> $$5.5 = 6.0 - (6.0 - 0)\frac{R1}{R1+R2}$$
>
> Both equations give the relationship that:
>
> $$\frac{0.5}{6.0} = \frac{R1}{R1+R2}$$
>
> There are two unknowns in this equation, so one of the resistor values must be chosen, and the other can be calculated.
>
> Let $R1 = 10k$ and then, from this, $R2 = 110k$.
>
> Most LEDs have a forward voltage drop of around 2 V and require approximately 10 mA for a reasonable light output. The value of $R7$ is given by:
>
> $$R7 = \frac{12 - 2.0}{10\,mA} = 1\,k\Omega$$

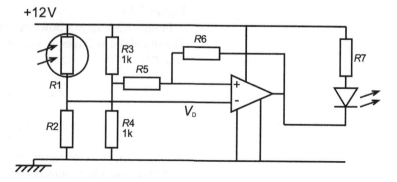

Figure 7.3.6

Figure 7.3.7

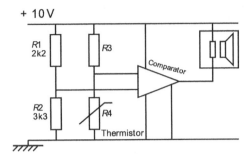

Problems 7.3.1

(1) The circuit of Figure 7.3.7 is to be used as a frost detector. The thermistor has a resistance of 1500 Ω at 0 °C. The alarm must sound when the thermistor resistance rises above 1500 Ω. In the circuit diagram:
   (a) Calculate a value for R3.
   (b) State which input is $V_{IN}+$ and which $V_{IN}-$.
   (c) Show how the circuit could be modified for 1.0 V hysteresis.

(2) A tachometer IC produces a voltage out which is proportional to the angular velocity of a sensor. The output is 1 V per 1000 revolutions per minute (rpm). Design a circuit which switches on a relay when the speed reaches 3800 rpm, and releases the relay when the speed falls to 3500 rpm.

(3) A fuel contents gauge is set up to produce an output of 1 V per 25 litres. Devise a comparator-based alarm circuit which will light a 'low fuel' LED when the contents fall below 10 litres. The LED should not go out again until the contents rise to 20 litres.

Window detector

A window detector is a circuit configuration which changes state when the input voltage is between two values. It may switch HIGH between the two values or it may switch LOW between the two values. The circuit can be built with two comparators as shown in Figure 7.3.8.

Figure 7.3.9 shows how the output switches as the input varies. The threshold voltages of V_H and V_L are set by the ratios of $R1$, $R2$ and $R3$:

$$V_L = V_S \frac{R3}{R1 + R2 + R3}$$

$$V_H = V_S \frac{R2 + R3}{R1 + R2 + R3}$$

$R6$ is the normal collector load of the output transistors of the comparators. The connection is called a **wired-or** because either of

Figure 7.3.8 *The window comparator*

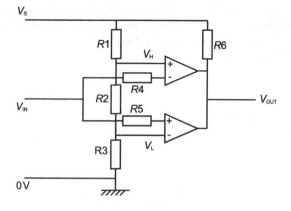

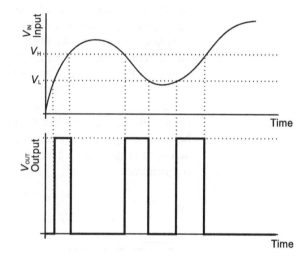

Figure 7.3.9 *The window comparator response*

the comparators can pull the output voltage down to ground. Resistors $R4$ and $R5$ are not critical in value and are typically in the range 1k–47k.

Problems 7.3.2

(1) If $R1 = 10k$, $R2 = 1k$, $R3 = 10k$ and the supply voltage is 12 V, calculate the window voltage range over which the output would be HIGH (12 V).
(2) Design a circuit which has an output which is normally at 0 V, but which switches to the +10 V supply whenever the input is between 6 V and 7 V.
(3) Design a circuit which has an output which is normally at the 15 V supply, but which switches down to 0 V whenever the input is between 1.4 V and 2.7 V.

7.4 Integrators

An integrator is a circuit which integrates (sums) the voltage at its input with time. This implies storing the signal which is present, and for this task a capacitor is used. The ideal integrator is shown in Figure 7.4.1.

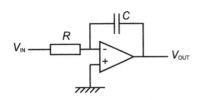

Figure 7.4.1 *The operational amplifier integrator*

When a constant voltage is applied to the input, the capacitor is charged by the constant current of V_{IN}/R. The full current balance is given by:

$$\frac{V_{IN}}{R} = -C\frac{dV_{OUT}}{dt} \quad \text{or} \quad V_{OUT} = -\frac{1}{RC}\int V_{IN}\,dt$$

so

$$V_{OUT} = -\frac{1}{RC}Vt$$

D.c. performance

Ideally, when the input voltage is 0 V, the output is zero. When a d.c. voltage is present at the input, the output ramps up – i.e. integrates the input with time. In the case of Figure 7.4.1, which is an inverting configuration, the input and output signals are as shown in Figure 7.4.2. This shows the steady d.c. input producing the integrated ramp output.

A practical op amp has offsets which cause the output to ramp to the supply rails even when no input signal is present. In this case, better biasing can be obtained by putting a large value resistor across the capacitor. This reduces the integration effect at very low frequencies.

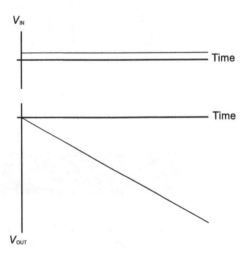

Figure 7.4.2 *Input and output voltages*

Example 7.4.1

What would be the rate of increase of V_{OUT} if $R = 10k$, $C = 1\,\mu F$ and a 100 mV signal is presented to the input?

$$V_{OUT} = -\frac{1}{RC}V_{IN}t$$

The rate of output ramp is given by:

$$\frac{V_{OUT}}{t} = -\frac{V_{IN}}{RCV_{IN}}\,V/s$$

$$= -\frac{100\,mV}{10k \times 1\,\mu F} = -10\,V/s$$

Clearly, it would not ramp down at −10 V per second for ever, but would stop (saturate) at the maximum output excursion of the operational amplifier. Commonly, this might be −13.5 V for an amplifier operating from +15 and −15 supplies. In this case, it would be after 1.35 seconds.

Problems 7.4.1

(1) In the circuit of Figure 7.4.4, calculate the rate of increase of the output voltage.

(2) If instead of a steady −10 mV signal, the circuit of Figure 7.4.4 is subjected to a square wave which switches from −10 mV to +10 mV, the output will be a triangle wave form as the output first ramps positive and then negative at the rate determined in question 1. When the input switches polarity, the current to the capacitor merely reverses direction and the output voltage ramps in the opposite direction. If the square wave has a frequency of 1 Hz, what will be the maximum amplitude of the resultant output waveform?

(3) In the circuit of Figure 7.4.4, sketch the output wave form if the input was a series of pulses which repeatedly rise to 1 V for 1 s and then fall to 0 V for 2 s.

Integrator applications

A sawtooth ramp generator can be constructed by periodically discharging the voltage across the capacitor. Transistor Q1 could be a bipolar transistor as shown in Figure 7.4.5 or a FET. The output is shown for a negative d.c. input voltage. A comparator could be used to detect when a certain threshold voltage had been reached and then turn the transistor ON. A triangle wave generator could

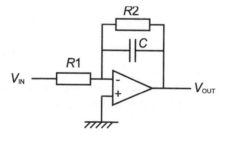

Figure 7.4.3 *Prevention of saturation at d.c.*

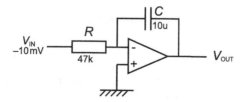

Figure 7.4.4

Figure 7.4.5 *The ramp generator*

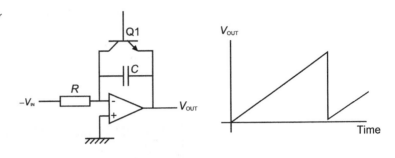

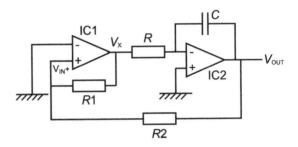

Figure 7.4.6 *A triangle wave generator*

be made by feeding back the output of the integrator to a comparator as in Figure 7.4.6.

IC2 is an integrator which operates in the manner discussed. IC1 is a comparator with positive feedback. At an instant when the comparator output is fully negative, the output of the integrator will ramp positive. This will continue until the voltage at $V_{IN}+$ exceeds 0 V. This will cause the comparator to switch and its output will go positive. The integrator will then start ramping down until V_{OUT} forces the voltage at $V_{IN}+$ to be less than 0 V. The comparator switches negative, and the process starts all over again. Analysing the current through R1 and R2:

$$\frac{V_X^+ - V_{OUT}}{R1 + R2} = \frac{V_X^+ - V_{IN}^+}{R2}$$

$V_{IN}+$ is zero, so:

$$V_X^+ \times R2 - V_{OUT} \times R2 = V_X^+ \times (R1 + R2)$$

$$-V_{OUT} \times R2 = V_X^+ \times R1$$

$$V_{OUT} = -\frac{R1}{R2} V_X^+$$

A similar proof will show that the opposite excursion of the integrator will produce:

$$V_{OUT} = -\frac{R1}{R2} V_X^-$$

The integration rate is:

$$\frac{V_X}{CR} \text{ V/s}$$

so the time to ramp up from the extremes of the integrator outputs would be:

$$T = \frac{CR}{V_X}\left(-\frac{R1}{R2}V_X^+ + \frac{R1}{R2}V_X^-\right)$$

$$= \frac{CR}{V_X}\left(\frac{R1}{R2}2V_X\right)$$

$$= 2CR\frac{R1}{R2}\text{ s}$$

The periodic time of oscillation is twice this, which makes the frequency:

$$f = \frac{1}{4CR}\frac{R2}{R1}$$

Example 7.4.2

In a circuit where $C = 10\,\text{nF}$, $R = 2.2\text{k}$, $R1 = 10\text{k}$, $R2 = 33\text{k}$:

(a) Calculate the frequency of oscillation.
(b) If the operational amplifiers are supplied from $\pm 15\,\text{V}$ so that their output saturation voltages are $\pm 13.5\,\text{V}$, what would be the amplitude of:
  (i) the square waveform;
  (ii) the triangle waveform.

(a)
$$f = \frac{1}{4CR}\frac{R2}{R1} = \frac{1}{4 \times 10 \times 10^{-9} \times 2.2 \times 10^3}\frac{33 \times 10^3}{10 \times 10^3}$$

$$= 37.5\,\text{kHz}$$

(b) (i) The square waveform switches between the positive and negative maximum output excursions, so the output square wave will be from $+13.5\,\text{V}$ to $-13.5\,\text{V}$.

(ii) The amplitude of the triangle wave is set by the resistors $R1$ and $R2$:

$$V_{\text{OUT}} = 13.5\,\text{V} \times \frac{R1}{R2}$$

$$= 13.5 \times \frac{10}{33}$$

$$= \pm 4.1\,\text{V}$$

Problems 7.4.2

(1) In the square-triangle oscillator of Figure 7.4.5, if $C = 15\,\text{nF}$, $R = 56\text{k}$, $R1 = 10\text{k}$ and $R2 = 20\text{k}$, find the frequency of oscillation and the maximum amplitude of the triangle wave when the operational amplifiers are supplied from $\pm 15\,\text{V}$.

(2) Design an oscillator which will output a 10 kHz triangle waveform of amplitude $\pm 10\,\text{V}$. Assume $\pm 15\,\text{V}$ supplies.

(3) Design an oscillator which will output a triangle waveform and square waveform, both of amplitude ±7.5 V. Assume ±15 V supplies. The square wave output will need attenuating in some way.
(4) Design an oscillator which will output a 440 Hz triangle and square waveform each of whose amplitude varies from +5 V to 0 V. Assume +15 V supplies. (One solution might be to use an operational amplifier with zero offset control to 'lift' ±2.5 V output to +5 V and 0 V.)

7.5 Active filters

A filter is a circuit which does not pass all frequencies equally. It could block low frequencies, high frequencies or a range of frequencies. The four main types of filter and their frequency response are shown in Figure 7.5.1:

- A low pass filter only passes low frequencies and blocks high frequencies.
- A high pass filter only passes high frequencies and blocks low frequencies.
- A band pass filter passes a range of frequencies.
- A band stop or notch filter blocks a range of frequencies.

An **active** filter is one which contains an active device. That is one which can amplify the original signal. With any filter, the features of interest are:

- the corner (or roll-off frequencies);
- The rate of roll-off.

The more complex the filter, the sharper the rate of roll-off (Figure 7.5.2).

The roll-off rate is always in multiples of -20 dB per decade. This means that if the corner frequency f_0 is 100 Hz, then:

- between 1000 Hz and 10 000 Hz (1 decade) the output would drop by 20 dB;

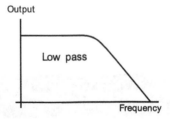
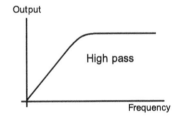
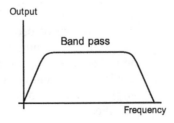
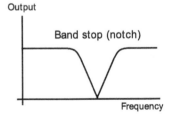

Figure 7.5.1 *Filter frequency responses*

Figure 7.5.2 *A filter response*

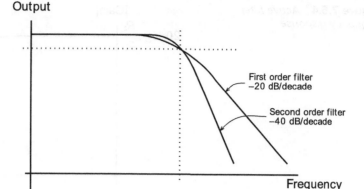

- between 10 000 Hz and 100 000 Hz (1 decade) the output would drop by 20 dB.

−20 dB per decade is numerically the same as −6 dB per octave. An octave is a doubling of frequency, so that:

- from 1000 Hz to 2000 Hz the output would drop by 6 dB;
- from 2000 Hz to 4000 Hz the output would drop by 6 dB;
- from 4000 Hz to 8000 Hz the output would drop by 6 dB; etc.

Inductors and capacitors are used in **passive filters** to determine the frequency response. In **active filters**, the use of inductors is most unusual. Capacitors are the preferred component because they are small, light and can be manufactured to a better tolerance than inductors for a given price. An ideal capacitor has an impedance of:

$$Z_C = \frac{1}{j\omega C} \Omega$$

where $\omega = 2\pi f$ radians/s.

A simple low pass active filter has a circuit diagram as shown in Figure 7.5.3.

The current flow through R_1 towards the −IN input is balanced by the current flow through the feedback components:

$$\frac{V_{IN}}{R_1} = -\frac{V_{OUT}}{R_2 \| C} \quad \text{where} \quad R_2 \| C = \frac{R_2 \, 1/j\omega C}{R_2 + 1/j\omega C} = \frac{R_2}{1 + j\omega C R_2}$$

$$= -\frac{V_{OUT}}{\dfrac{R_2}{1 + j\omega C R_2}}$$

from which:

$$\frac{V_{OUT}}{V_{IN}} = -\frac{R_2}{R_1(1 + j\omega C R_2)}$$

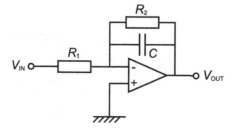

Figure 7.5.3 *A simple active low pass filter*

Figure 7.5.4 *Active filter frequency response*

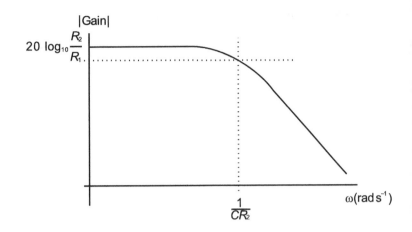

At low frequencies, when $\omega = 0$:

$$\frac{V_{OUT}}{V_{IN}} = -\frac{R_2}{R_1}$$

which is the same as for a conventional inverting amplifier.

At high frequencies, when $\omega CR_2 \gg 1$, the expression simplifies to:

$$\left|\frac{V_{OUT}}{V_{IN}}\right| = \frac{R_2}{\omega CR_1 R_2}$$

As ω increases, V_{OUT}/V_{IN} gets smaller, and the gain is seen to roll off with larger and larger frequencies. The transition occurs when $\omega CR_2 = 1$, i.e. the corner frequency is given by:

$$\omega = \frac{1}{CR_2}$$

(see Figure 7.5.4).

A high pass filter would result from the circuit of Figure 7.5.5:

$$\frac{V_{IN}}{R_1 + 1/j\omega C} = -\frac{V_{OUT}}{R_2}$$

$$\frac{V_{OUT}}{V_{IN}} = -\frac{R_2}{R_1 + 1/j\omega C} = -\frac{j\omega CR_2}{1 + j\omega CR_1}$$

Is this the **transfer function** of a high pass filter? Consider its response at low and high frequencies. When ω is very large, $j\omega CR_1 \gg 1$:

$$\frac{V_{OUT}}{V_{IN}} = -\frac{j\omega CR_2}{j\omega CR_1} = -\frac{R_2}{R_1}$$

The gain is determined by the resistor ratio, like in a conventional inverting amplifier.

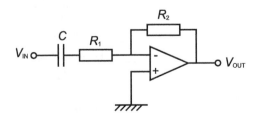

Figure 7.5.5 *An active high pass filter*

Figure 7.5.6 *High pass filter frequency response*

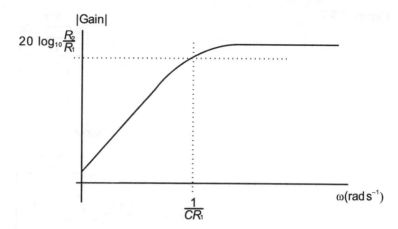

When ω is very small, $j\omega CR_1 \ll 1$:

$$\frac{V_{OUT}}{V_{IN}} = -j\omega CR_2$$

As ω gets smaller, V_{OUT}/V_{IN} gets smaller. The transition occurs when $\omega CR_1 = 1$ (Figure 7.5.6).

Problems 7.5.1

(1) In the filters of Figure 7.5.7, calculate the low frequency gain and the roll-off frequencies.
(2) In the filters of Figure 7.5.8, calculate the high frequency gain and the roll-off frequencies.
(3) Design a low pass filter to give a low frequency gain of -25 and a roll-off frequency of 1.5 kHz.
(4) Design a high pass filter to give a high frequency gain of -6 and a roll-off frequency of 3.4 kHz.

Second order 'Sallen and Key' filters

Active filters come in many topologies, and a proper coverage would be the subject of a book in itself. This small topic of active filters as a subsection of operational amplifiers gives only the basic introduction to the range and variety that are available. So far, we have only looked at first order filters. These have a roll-off rate which has limited use in most practical situations. We shall now look at an industry-standard set of filters which were developed by Sallen and Key and can be configured into low pass and high pass variations. They can even be set up for first order band pass functions.

Figure 7.5.9 shows the fundamental layout for a low pass filter. It follows the general rule that for each order of operation, there should be an RC time constant in the circuit. So, in this case, there are two RC time constants which hints towards second order operation. The full proof of the transfer function is:

Figure 7.5.7

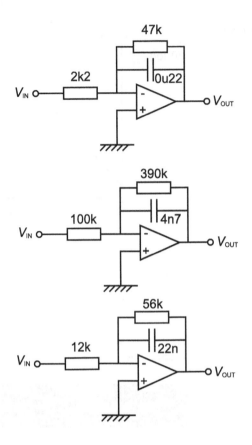

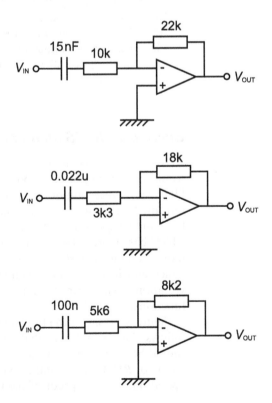

Figure 7.5.8

Figure 7.5.9 *A Sallen and Key low pass filter*

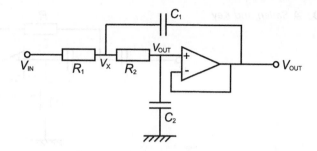

$$\frac{V_{OUT} - V_X}{1/j\omega C_1} + \frac{V_{IN} - V_X}{R_1} = \frac{V_X - V_{OUT}}{R_2}$$

Multiply throughout by $R_1 R_2$ to find:

$$R_1 R_2 V_{OUT} j\omega C_1 - j\omega C_1 V_X R_1 R_2 + V_{IN} R_2$$
$$- V_X R_2 = V_X R_1 - V_{OUT} R_1$$

Collecting terms in V_{OUT}, V_X and V_{IN} results in:

$$V_{OUT} R_1 (1 + j\omega C_1 R_2) + V_{IN} R_2 = V_X (R_1 + R_2 + j\omega C_1 R_1 R_2)$$

Now, examining the voltage drop across R_2 and C_2 results in the relationship between V_X and V_{OUT}:

$$V_{OUT} = \frac{V_X \, 1/j\omega C_2}{R_2 + 1/j\omega C_2} = V_X \frac{1}{1 + j\omega C_2 R_2}$$

Substituting for V_X in the previous equation results in a relationship between V_{IN} and V_{OUT}:

$$\frac{V_{OUT}}{V_{IN}} = \frac{1}{1 + j\omega(R_1 C_2 + R_2 C_2) + j^2 \omega^2 (R_1 R_2 C_1 C_2)}$$

When ω is very small, the equation reduces to:

$$\frac{V_{OUT}}{V_{IN}} = 1$$

When ω is very large, the ω^2 terms dominate the denominator and the equation simplifies to:

$$\frac{V_{OUT}}{V_{IN}} = \frac{1}{j^2 \omega^2 R_1 R_2 C_1 C_2}$$

The j^2 imparts a 180° phase shift, and the amplitude falls with the square of the frequency. This gives it its -40 dB per decade roll-off. The roll-off frequency occurs when the real terms of the denominator cancel, i.e. when $1 = \omega^2 R_1 R_2 C_1 C_2$. This occurs at a frequency of:

$$f = \frac{1}{2\pi \sqrt{R_1 R_2 C_1 C_2}} \text{ Hz}$$

In many applications, $R_1 = R_2 = R$ and $C_1 = C_2 = C$, so:

$$f = \frac{1}{2\pi RC} \text{ Hz}$$

Swapping the positions of resistors and capacitors results in a high pass filter with the same roll-off frequency (Figure 7.5.10).

Figure 7.5.10 *A Sallen and Key high pass filter*

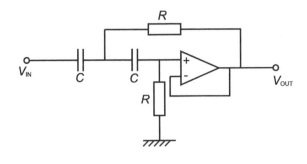

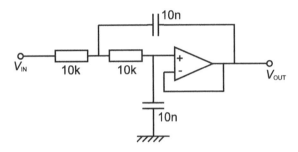

Figure 7.5.11

> ### Example 7.5.1
>
> Calculate the roll-off frequency of the filter shown in Figure 7.5.11. What is the gain at **four** times this frequency?
>
> The gain is given by:
>
> $$f = \frac{1}{2\pi RC} = \frac{1}{2 \times \pi \times 10 \times 10^3 \times 10 \times 10^{-9}} = 1.6\,\text{kHz}$$
>
> This is a second order filter, so the gain rolls off at $-12\,\text{dB}$ per octave. At 3.2 kHz, the gain will be $-12\,\text{dB}$. At 6.4 kHz, the gain will be $-24\,\text{dB}$ or for 1 V input the output will be 0.063 V.
>
> To find the output at any other frequency, the full V_{OUT}/V_{IN} equation would have to be used.

Problems 7.5.2

(1) In the low pass filter of Figure 7.5.12(a) calculate the roll-off frequency.
(2) In the high pass filter of Figure 7.5.12(b) calculate the roll-off frequency.
(3) Design a Sallen and Key low pass filter to roll off at 100 Hz.
(4) Design a Sallen and Key high pass filter to roll off at 5 kHz.
(5) Design a Sallen and Key low pass filter to reduce an input signal by $-40\,\text{dB}$ at 50 Hz.

Figure 7.5.12

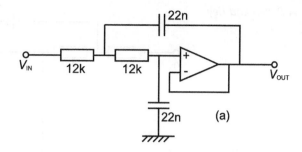

(a)

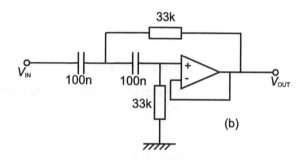

(b)

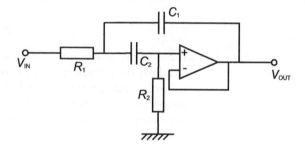

Figure 7.5.13 *A Sallen and Key band pass filter*

A Sallen and Key band pass filter would be constructed as in the circuit of Figure 7.5.13. In this case, **two** of the resistor/capacitor pairs have been transposed and the general case of both resistors and both capacitors being equal does **not** hold. The value of the R_1-C_1-R_2-C_2 pairs determines the low pass and high pass frequencies. The roll-off is only -20 dB at either end. This configuration is only rarely used because designers invariably want sharp cut-offs to filters.

A better band pass response can be achieved by cascading a second order low pass and a second order high pass filter. This does use two op amps, but packages such as the TL072 boast two operational amplifiers in a single 8-pin DIL arrangement.

Cascading two second order low pass filters would result in a fourth order low pass filter with a roll-off rate of -80 dB per decade. So a fourth order band pass filter could be made from a single quad op amp package such as the LM324.

Sallen and Key filters with gain

Gain can be introduced into a Sallen and Key filter by modifying the feedback as in Figure 7.5.14. The feedback components in this low pass filter introduce a gain term of:

Figure 7.5.14 *A Sallen and Key filter with gain*

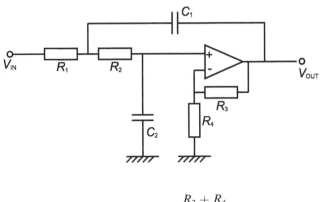

$$A_V = \frac{R_3 + R_4}{R_4}$$

Notch filters

A notch filter could be made from the output of a cascaded low pass and high pass filter. However, most practical notch filters use the twin-T filter as shown in Figure 7.5.15.

The twin-T is like a parallel connected low pass and high pass filter. When the resistors and capacitors are selected in the ratios shown, the notch occurs at a frequency of:

$$f = \frac{1}{2\pi RC} \text{Hz}$$

It has a very sharp notch (infinite attenuation in theory) but is hard to adjust since three resistors must be changed simultaneously. The resistors and capacitors must be matched quite closely in value, temperature coefficient and stability if the performance is to approach anything like its theoretical performance. A common variation is to feed back a proportion of the output into the bottom of the T (see Figure 7.5.16). This gives control of the Q factor of the filter (sharpness of the notch).

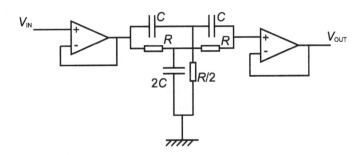

Figure 7.5.15 *The twin-T notch filter*

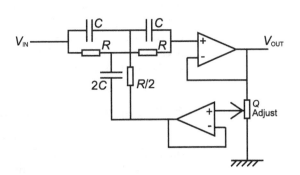

Figure 7.5.16 *Adjustable Q notch filter*

7.6 Transconductance amplifiers

So far we have looked at the operational amplifier as a voltage controlled voltage source. There is no reason why the device should not respond to a current input or why it should not provide a current output. One interesting range of devices is the transconductance amplifier, sometimes called the OTA or operational transconductance amplifier. This produces a current output proportional to the voltage input. The gain of amps output for volts input has the units of conductance and uses the symbol G.

An early, but still useful, OTA is the CA3080:

Supply voltage	± 2 V to ± 15 V
Input resistance	26 k
Slew rate	50 V/μs
Transconductance	10 000 μs (μA/V)
Output resistance	15 M

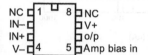

Figure 7.6.1 *The CA3080*

When a load resistor R is connected to the output of the OTA, response is simply $V_{OUT} = G \times R \times (V_{IN}^+ - V_{IN}^-)$ The pin out of the CA3080 is shown in Figure 7.6.1.

Most practical OTAs have a version of the control pin 5 'amp bias' which allows a variation of the transconductance G. It connects to the common emitter section of the input long tailed pair. This is shown in the internal schematic of Figure 7.6.2.

The control of current by Q3 in the long tailed pair formed by Q1 and Q2 varies the transconductance of the amplifier. With this third input, you create a gain-controlled amplifier, and the possibility of building circuits such as voltage multipliers. Pin 5 is a current input, so care must be taken to limit that current when operating from a voltage source. Always use an input resistor into pin 5 when driving from a voltage source!

Figure 7.6.3 shows the effect which the variation of control current has on the gain (transconductance) of the amplifier.

The current output feature of an OTA means that it could be directly connected to the output of other OTAs. Current mixing would take place into a common load. This implies that the device should be useful for applications where signal summing takes place after amplification or signal conditioning.

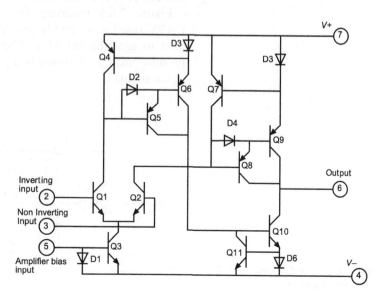

Figure 7.6.2 *Internal schematic of the CA3080*

Figure 7.6.3 *Variation of transconductance with bias current*

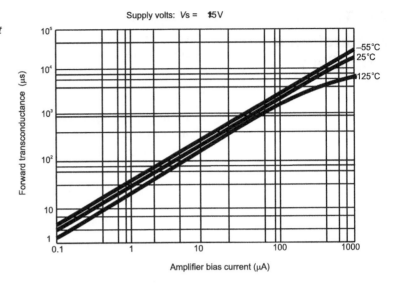

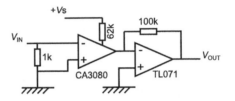

Figure 7.6.4 *OTA/op amp pair*

The current output can drive into a resistive load to create a voltage output, or if more (voltage) amplification is required, the OTA will directly drive a conventional operational amplifier as in Figure 7.6.4.

In this circuit, the output voltage of the op amp is given by $-I_{OUT}/R_F$. I_{BIAS} can be calculated when the supply voltage is known; the graph of Figure 7.6.3 will give the approximate transconductance; the CA3080 output current is calculated from the transconductance and the CA3080 input voltage; the output voltage of the TL071 is calculated from the CA3080 output current and the 100k feedback resistor. Can you derive a formula for this?

Figure 7.6.5 (courtesy of Harris Semiconductor) shows a CA3080 used as a sample and hold amplifier. The bias control is used to gate the amplifier. These devices are also often used to produce an output which is a product of the bias input and the signal input.

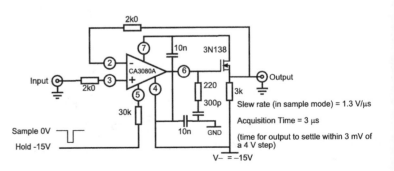

Figure 7.6.5 *A CA3080 sample and hold amplifier*

8 Oscillators

Summary

Oscillators are devices which produce a regular repeatable waveform as an output. In this chapter the study has been split into two categories: sinewave oscillators and non-sinewave oscillators (sometimes called more simply waveform generators). By the end of this chapter, you will have some understanding of the operating principles of oscillators and the techniques needed to design them.

A sine wave is one unique frequency. Other waveforms can be synthesized by adding combinations of sine waves with particular frequency, amplitude and phase relationships. For instance, a square wave can be made by adding together **odd** harmonics to the fundamental frequency. However, the sine wave is the 'purest' form of oscillator. This is not to mean that sinewave oscillators are the simplest or easiest to design or build. Imperfections in component matching or drift of the values with time or temperature usually result in distortions in the output – additional frequency signals may be present other than the one intended. A sinusoidal oscillation usually results because the circuit has components which have a particular response at **one frequency only**. This attribute is used as a feature to make the circuit boost or select that frequency.

8.1 The tuned collector oscillator

The parallel connected LC circuit will resonate at one fundamental frequency given approximately by:

$$f = \frac{1}{2\pi\sqrt{LC}} \text{Hz}$$

If the input amplitude is kept constant, the output voltage of the circuit of Figure 8.1.1 will vary with frequency. At the resonant frequency, the output will be at a maximum as shown. The width

Figure 8.1.1 *Resonant output stage*

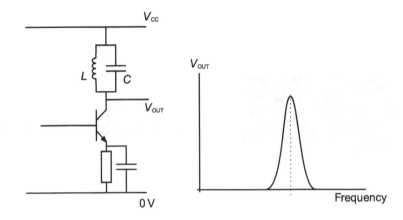

(= bandwidth) of the output response is governed by the resistance of the windings of the coil. The lower the resistance, the narrower the response. This situation is usually referred to as having a higher selectivity since only a very narrow band of frequencies will be amplified.

The gain of a common emitter bipolar transistor amplifier is proportional to the collector resistance. In the situation of Figure 8.1.1, the impedance of the *LC* collector load is related to the signal frequency. At resonance, the **dynamic resistance**, as it is called, is at a maximum and is given by the formula:

$$R_D = \frac{L}{Cr}$$

where

R_D is the dynamic resistance
L is the inductance
r is the coil resistance
C is the capacitance

This circuit has an excellent gain response at one frequency only. If the output is connected back to the input with the correct phase relationship, the circuit might be induced to oscillate at the resonant frequency.

Figure 8.1.2 shows a BJT and FET version of the tuned collector/drain oscillator. The dots on the transformer are the convention which is adopted to show the polarity of the windings. In this instance, the transistors form an inverting amplifier, so the transformers must be connected as shown. The transformer will have a turns ratio between the two sets of windings, so if the gain is to be set correctly for oscillation to take place, then a form of gain adjustment must be available.

The small circuit modification where R_E or R_S is replaced with a variable resistor allows the gain of each tuned amplifier to be set so that a pure sine wave will result. This would always be needed in practice because the h_{FE} or g_m of the transistors would not be known in advance.

Another version of the circuit uses a long tailed pair of transistors as shown in Figure 8.1.3. Feedback is capacitively fed by *C*2 from the tuned collector circuit to the base of the other transistor Q1 in the differential amplifier. In this way, the 180° phase shift is obtained.

Figure 8.1.2 The tuned collector and drain oscillator with transformer feedback

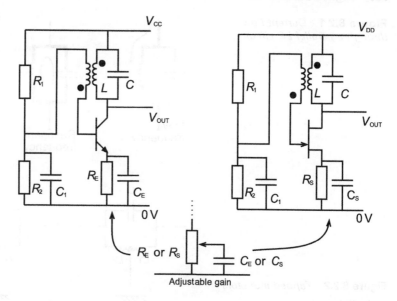

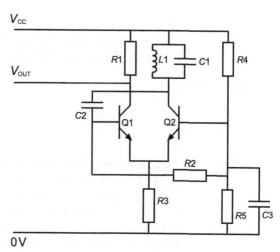

Figure 8.1.3 A tuned collector oscillator with a long tailed pair

Bias for Q2 is provided by $R4$ and $R5$ while Q1 picks up its bias from an additional resistor $R2$. The frequency is set by the tuned pair $L_1 C_1$ at:

$$f = \frac{1}{2\pi\sqrt{L_1 C_1}} \text{ Hz}$$

8.2 Colpitts and Hartley oscillators

The parallel tuned LC circuit will resonate at a frequency of approximately:

$$f = \frac{1}{2\pi\sqrt{LC}}$$

Energy oscillates between the inductor and capacitor. The inductor stores the energy in a magnetic field, and the capacitor in an electric field. At resonance, the impedance is at a maximum, and little current is drawn from the supply. The Radio Society of Great Britain (RSGB) Handbook gives a good visualization of the current drawn by a resonant circuit.

Figure 8.2.1 *Current flow through a parallel LC circuit*

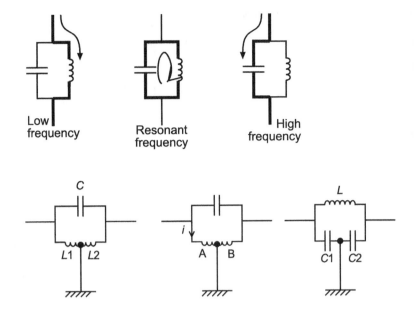

Figure 8.2.2 *Tapped inductors and capacitors*

Now consider the effect of tapping the inductor part way along its length. The circuit still resonates at a frequency of:

$$f = \frac{1}{2\pi\sqrt{LC}}$$

Now, however, the voltage across the inductor 'see-saws' about the earthed tap. When the current I passes through the inductor in the direction shown, then end A will be positive while end B will be negative with respect to the tap. This forms a basic Hartley circuit. A Colpitts circuit is formed by tapping the capacitor into two parts, $C1$ and $C2$. Since they are in series, the total capacitance will be:

$$\frac{1}{C_T} = \frac{1}{C_1} + \frac{1}{C_2}$$

In both of the circuits of Figure 8.2.3, the voltage BE is out of phase with the voltage CE. This is one of the requirements for an inverting amplifier to oscillate. The other is that the amplifier should amplify the signal by at least the amount given by V_{CE}/V_{BE}:

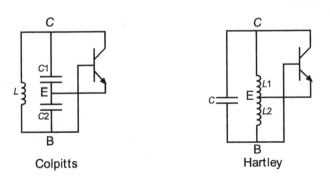

Figure 8.2.3 *BJT Colpitts and Hartley oscillators*

Oscillators

	Colpitts	Hartley
Gain	$= \dfrac{V_{CE}}{V_{BE}}$	$= \dfrac{V_{CE}}{V_{BE}}$
	$= \dfrac{i \times X_{C1}}{i \times X_{C2}}$	$= \dfrac{i \times X_{L1}}{i \times X_{L2}}$
	$= \dfrac{1}{\omega C_1} \times \dfrac{\omega C_2}{1}$	$= \dfrac{\omega L_1}{\omega L_2}$
	$= \dfrac{C_2}{C_1}$	$= \dfrac{L_1}{L_2}$
Frequency	$\dfrac{1}{2\pi\sqrt{L\dfrac{C1\,C2}{C1+C2}}}$	$\dfrac{1}{2\pi\sqrt{(L1+L2)C}}$

Colpitts oscillators

In the series-fed case (Figure 8.2.4), the quiescent collector voltage is equal to the supply voltage. The **maximum** output voltage swing is from 1 V to 17 V. The shunt-fed oscillator has a quiescent collector voltage of about 4.5 V. The maximum output voltage swing is 8 V pk-pk.

Typically $C2$ may be 10 to 30 times the size of $C1$. This ensures that there is enough gain in the transistor stage to force positive feedback to start oscillation.

Capacitor $C2$ can be used to generate a.c. signals between audio frequencies and a few GHz. The junction between the two capacitors is directly grounded in one case and connected via a low resistance in the other.

Hartley oscillators

In Figure 8.2.5, the tap on the inductor is connected to the $+V_{CC}$ line. As far as the a.c. is concerned, this is the same as connecting to ground (since any power supply has a low internal resistance and hence low impedance to a.c. signals).

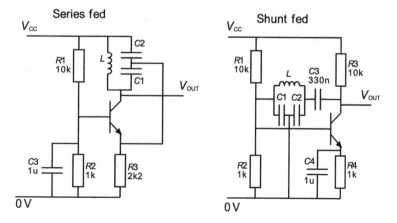

Figure 8.2.4 *Series and shunt Colpitts oscillators*

Figure 8.2.5 *Hartley oscillator*

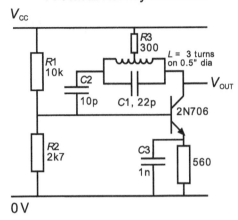

100 MHz Hartley oscillator

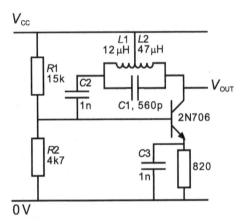

Figure 8.2.6 *A Hartley oscillator*

Examples 8.2.1

In the circuit of Figure 8.2.6, calculate the frequency of oscillation. What controls the gain of the amplifier?

In the circuit, the formula for resonant frequency is:

$$f = \frac{1}{2\pi\sqrt{(L1 + L2)C}} \text{ Hz}$$

where

$$L1 = 12\,\mu\text{H}$$
$$L2 = 47\,\mu\text{H}$$
$$C = 560\text{p}$$

So $f = 875\,\text{kHz}$.

The gain of the amplifier is set by the parameters of the transistor and the value of the dynamic resistance of the tuned circuit. This will be affected by the value of L and C and the resistance of the coil.

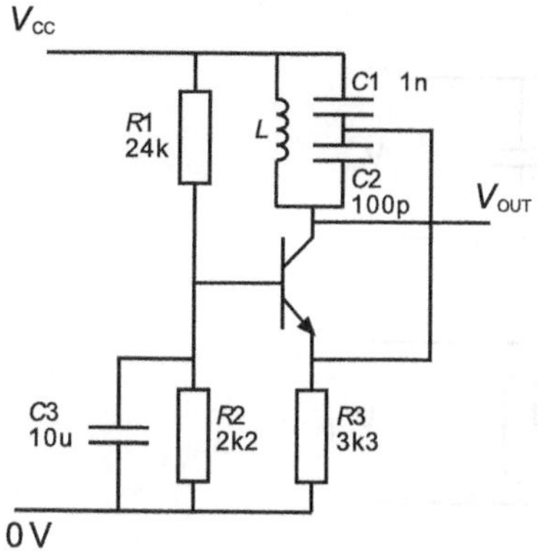

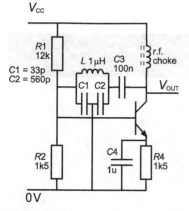

Figure 8.2.7

Figure 8.2.8

Problems 8.2.1

(1) The formula for an air wound inductor of length b'' and diameter a'' with n turns is given approximately by:

$$L = \frac{a^2 n^2}{5(3a + 9b)} \mu H$$

Such an inductor is made by winding 900 turns on a former 0.5″ long and 0.2″ diameter. It is to be included in a tuned collector circuit with the intention of producing oscillations at 300 kHz.
 (a) Calculate the value of the capacitor required.
 (b) If the amplifier gain is −150, calculate the secondary turns required on the feedback transformer.

(2) In Figure 8.2.7, calculate the frequency of oscillation and the minimum gain needed from the amplifier for oscillation to take place.

(3) Calculate the expected frequency of oscillation of the circuit of Figure 8.2.8.

8.3 RC sinewave oscillators

LC oscillators work because of the resonance characteristic between the inductor and capacitor. This section deals with RC oscillators. The RC network has to have a particular property at one frequency only if the output is to be a sine wave.

Phase shift ladder

An inverting amplifier has a phase shift of 180° for sine waves. To provide positive feedback, the output signal must also be phase shifted by 180° if oscillation is to take place. A capacitor exhibits 90° phase shift between current and voltage. An RC circuit could produce, in theory, almost 90° phase shift between input and output signals, but in practice three stages of RC network are used as in Figure 8.3.1.

Each of the stages introduces an attenuation and phase shift of an applied a.c. signal which depends on the frequency of that signal. There is one frequency at which the phase shift is exactly 180°. This occurs at:

$$f = \frac{\sqrt{6}}{2\pi RC} \text{ Hz}$$

At this frequency, the output amplitude would be:

$$V_{OUT} = -\frac{V_{IN}}{29}$$

Figure 8.3.1 *An RC ladder network*

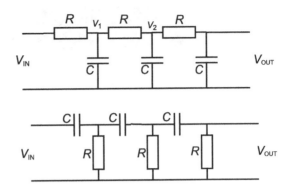

The minus sign of course indicates a phase shift of 180°.

Mathematics in action

Summing the current at node $V1$:

$$\frac{V_{IN} - V1}{R} - \frac{V1}{X_C} - \frac{V1 - V2}{R} = 0$$

$$\frac{V_{IN}}{R} - V1\left(\frac{2}{R} + \frac{1}{X_C}\right) + \frac{V2}{R} = 0$$

Summing the current at node $V2$:

$$\frac{V1 - V2}{R} - \frac{V2}{X_C} - \frac{V3 - V_{OUT}}{R} = 0$$

$$\frac{V1}{R} - V2\left(\frac{2}{R} + \frac{1}{X_C}\right) + \frac{V_{OUT}}{R} = 0$$

Summing the current at node V_{OUT}:

$$\frac{V2 - V_{OUT}}{R} = \frac{V_{OUT}}{X_C} \quad \frac{V2}{R} - V_{OUT}\left(\frac{1}{R} + \frac{1}{X_C}\right) = 0$$

From node $V2$:

$$V2 = V_{OUT} R\left(\frac{1}{R} + \frac{1}{X_C}\right)$$

From node $V1$:

$$V1 = V2 R\left(\frac{2}{R} + \frac{1}{X_C}\right) - V_{OUT}$$

$$= V_{OUT} R^2 \left(\frac{1}{R} + \frac{1}{X_C}\right)\left(\frac{2}{R} + \frac{1}{X_C}\right) - V_{OUT}$$

Substituting this into the node 1 relationship, and replacing X_C with $1/j\omega C$ results in:

$$V_{IN} = V_{OUT}(1 + 6j\omega CR + 5j^2\omega^2 C^2 R^2 + j^3\omega^3 C^3 R^3)$$

$$V_{IN} = V_{OUT}(1 + 6j\omega CR - 5\omega^2 C^2 R^2 - j\omega^3 C^3 R^3)$$

When $6j\omega CR - j\omega C^3 R^3 = 0$ then $6 = \omega^2 C^2 R^2$ or:

At this frequency:

$$\omega = \frac{\sqrt{6}}{CR} \qquad f = \frac{\sqrt{6}}{2\pi CR}$$

$$V_{IN} = V_{OUT}\left(1 - \frac{5 \times 6 \times C^2 R^2}{C^2 R^2}\right)$$

or

$$\frac{V_{OUT}}{V_{IN}} = -\frac{1}{29}$$

The network of Figure 8.3.1 can be incorporated into amplifier circuits in several different ways. Figure 8.3.2 shows the oscillator. $R1$ and $R2$ set the BJT bias in the usual way. R_E is a variable resistor which sets the gain to 29. The three RC stages perform a current division, feeding the signal back to the base. There are two main drawbacks:

(i) Varying the frequency of oscillation over a significant range means simultaneously altering three resistors (or capacitors).
(ii) To obtain a true sinusoidal output, the positive feedback must be limited to that which is sufficient just to maintain oscillation. This is not so easily done with this type of circuit.

An FET could be used as the active element, since in its basic common source layout it will provide the inverting amplifier which is needed by this network (Figure 8.3.3).

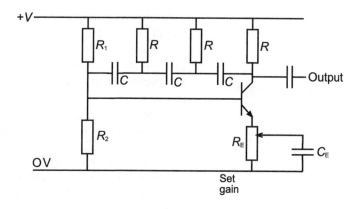

Figure 8.3.2 *BJT-based RC ladder oscillator*

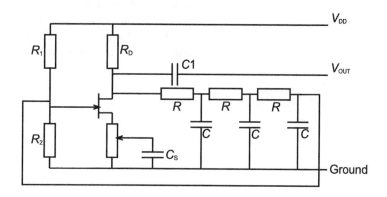

Figure 8.3.3 *An FET-based RC ladder oscillator*

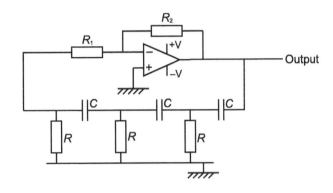

Figure 8.3.4 *An operational amplifier based RC oscillator*

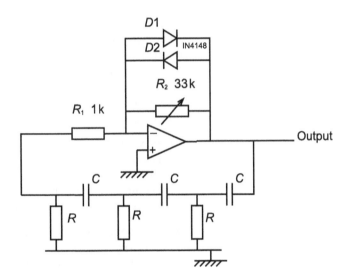

Figure 8.3.5 *Amplitude control*

An operational amplifier can also be used to act as the inverting amplifier with the gain of −29 (Figure 8.3.4). R2 and R1 would be chosen to be in the ratio of 29:1. More likely, there would be other gain controlling devices which would limit the gain to 29 to ensure that oscillation takes place. Figure 8.3.5 shows one way of achieving this effect.

In this circuit, R2 and R1 are chosen to give a maximum gain of −33. The diodes prevent the voltage across R2 from rising above 0.6 V. As soon as it tries to do so, the diodes conduct and the resistance from the output to the inverting input falls. When the amplifier gain falls, the oscillations will start to cease, but as soon as the voltage across R2 falls below 0.6 V, the diodes stop conducting and the amplifier gain once again rises. The signal feedback thus limits the gain of the amplifier and keeps the oscillator oscillating without saturation. There is a small amount of distortion caused by the diode characteristic. The disadvantage of this circuit is the low output amplitude of around 0.6 V.

8.4 Wien Bridge oscillators

The Wien Bridge is another type of RC network which has specific amplitude/phase characteristics at one frequency only (Figure 8.4.1).

At a frequency of:

$$f = \frac{1}{2\pi RC}$$

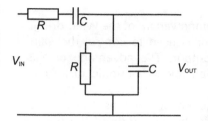

Figure 8.4.1 *The Wien Bridge*

the network has a characteristic of:

$$\frac{V_{OUT}}{V_{IN}} = \frac{1}{3}$$

Mathematics in action

$$V_{OUT} = V_{IN} \times \frac{R \| X_C}{R \| X_C + R + X_C}$$

$$= V_{IN} \times \frac{\left(\frac{R \times 1/j\omega C}{R + 1/j\omega C}\right)}{\left(\frac{R \times 1/j\omega C}{R + 1/j\omega C}\right) + R + 1/j\omega C}$$

With patience, this can be simplified down to:

$$\frac{V_{OUT}}{V_{IN}} = \frac{j\omega CR}{1 - \omega^2 C^2 R^2 + 3j\omega CR}$$

Zero phase shift occurs when:

$$1 - \omega^2 C^2 R^2 = 0$$

That is, when:

$$w = \frac{1}{RC} \quad \text{or} \quad f = \frac{1}{2\pi RC}$$

The gain at this frequency is:

$$\frac{V_{OUT}}{V_{IN}} = \frac{j\omega CR}{3j\omega CR} = \frac{1}{3}$$

To make an oscillator, the network must be part of an amplifier with a gain of +3. This can work very well with discrete transistors, but the usual configuration is with operational amplifiers. Figure 8.4.2 shows the basic circuit with the gain controlling diodes D1 and D2.

In this circuit VR1 and R1 form the negative feedback network with a maximum gain of $20/4.7 = 4.25$. If the amplifier was allowed to retain this high level of gain, the output peaks of the sinusoid would saturate and a very distorted sine wave would result. As soon as the output rises to an amplitude large enough to make the diodes conduct, the low forward resistance of the diodes reduce the gain of the amplifier. The output amplitude will fall until the diodes are on the point of conducting ($\sim 0.6\,V$). The RC network connects the output to the non-inverting input to provide the positive feedback. Since the network only has a 0° phase feedback at a frequency of $f = \frac{1}{2\pi RC}$ then this is the frequency at which it will oscillate.

If the output is still too small, then the version of the circuit in Figure 8.4.3 could be used. Here, two Zener diodes are used to limit the output voltage to (in this case) $3.3\,V + 0.6\,V = 3.9\,V$ pk-pk approximately.

Figure 8.4.4 shows a common variant of the Wien Bridge oscillator where a thermistor is used to control the gain, and hence stabilizes the output. The advantage of this circuit is that it has a much lower harmonic distortion. Values better than 0.1% are easily achievable. The thermistor has a negative temperature coefficient: as it warms up, the resistance falls. Thus, as the amplitude of the output increases, the negative (gain controlling) feedback resistor causes the gain to drop. This reduces the output amplitude and stabilizes the oscillator.

The value of $R1$ depends on the ambient value of resistance of the thermistor. $R1$ should be slightly less than one-half of the value of the thermistor to give an overall gain of 3. In practice, of course, $R1$ would be a variable pre-set resistor.

To control the frequency of the oscillator, a ganged potentiometer would be used (Figure 8.4.5). This would simultaneously vary both of the Rs in the Wien Bridge. The fixed resistors in series with the variable element prevent the output being shorted to ground.

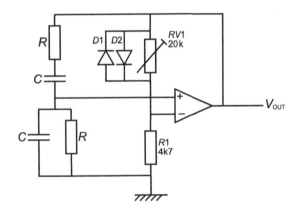

Figure 8.4.2 Wien Bridge oscillator

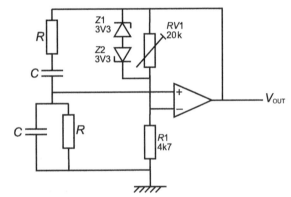

Figure 8.4.3 Wien Bridge oscillator with increased output amplitude

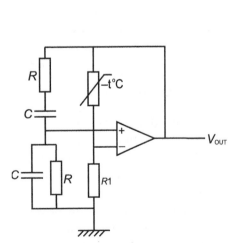

Figure 8.4.4 Thermistor stabilized Wien Bridge oscillator

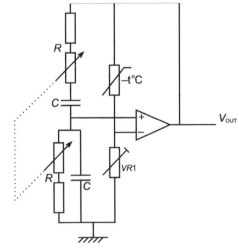

Figure 8.4.5 Frequency control of the Wien Bridge oscillator

Figure 8.5.1 *The twin-T network*

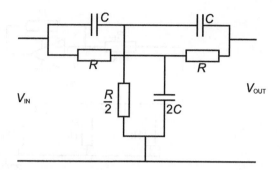

8.5 Twin T oscillators

The twin T configuration has the property that at one frequency $\left(\text{given by } f = \dfrac{1}{2\pi RC}\right)$ it is a perfect attenuator: it blocks all signals (Figure 8.5.1).

If used in the feedback path of an amplifier, then at this frequency it will have infinite gain. In practice, however, mismatches between the components mean that the attenuation is finite. Figure 8.5.2 shows a simple 1 kHz oscillator.

Example 8.5.1

In the circuit shown in Figure 8.5.3, calculate:

(a) the frequency of oscillation;
(b) the ideal value of R2 for oscillation to start.

(a)
$$f = \frac{\sqrt{6}}{2\pi RC} = \frac{\sqrt{6}}{2 \times \pi \times 1k \times 100n} = 3.898 \text{ kHz}$$

(b) The theoretical gain $= -29$:

$$R1 = 10k$$
$$R2 = 29 \times 10k = 290k$$

Note that in this configuration, $R1 \gg R$. This is to prevent significant current flowing out of the last RC node towards the op amp. If it did so, the accuracy of the design equation would be compromised.

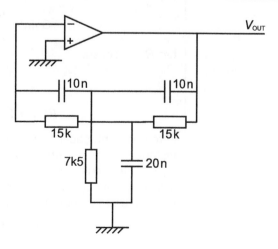

Figure 8.5.2 *Twin-T sinusoidal oscillator*

Figure 8.5.3

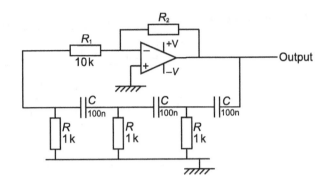

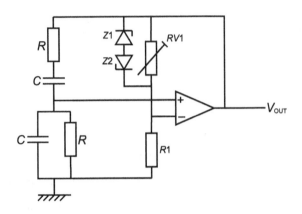

Figure 8.5.4

Example 8.5.2

Design a 440 Hz sinewave oscillator based on the circuit of Figure 8.5.4. The output should be 1.5 V_{RMS}.

(a) Frequency determining components:

$$f = \frac{1}{2\pi RC}$$

Let $C = 10$ nF. Then:

$$R = \frac{1}{2 \times \pi \times 440 \times 10 \times 10^{-9}} = 36\,k\Omega$$

(b) Gain control:

$$\text{Gain} = \frac{R1 + RV1}{R1}$$

Let $R1 = 1$k, so:

$$3 = \frac{1 + RV1}{1}$$

$$RV1 = 3 - 1 = 2\,k\Omega$$

Often, it is practical to use just the calculated value. In this instance, most designers would create $RV1$ from a variable resistor in series with a fixed resistor. This prevents the resistance falling to zero ohms when the wiper is moved to the one end of the track. A good compromise might be:

$$RV1 = 1.5 \text{ fixed} + 1\text{k variable}$$

This gives a gain range of 2.5 to 3.5:

$$1.5\,V_{RMS} = 4.24\,V_{pk-pk}$$

= two 3.6 V Zener diodes back to back (i.e. 3.6 V reverse breakdown in series with 0.6 V forward voltage drop).

Problems 8.5.1

(1) In *Figure 8.5.5*:
   (a) At what frequency will this *RC* ladder circuit oscillate?
   (b) Why is it advisable to place a buffer amplifier at point X?

(2) In *Figure 8.5.6*:
   (a) Calculate the frequency of oscillation of the circuit.
   (b) What percentage of R_E should the wiper be set to in order that the amplifier has a gain of -29?

(3) Design an op-amp-based *RC* ladder circuit which will produce a 1760 Hz sinewave output.

(4) In *Figure 8.5.7*:
   (a) At what frequency will the circuit oscillate?
   (b) What is the minimum value of *R2*?

(5) In *Figure 8.5.8*:
   (a) Calculate the expected frequency of oscillation.
   (b) If the thermistor has a resistance of 5k at 10 °C, what value of *R1* would you recommend?
   (c) Would you expect the oscillation to stop working at:
      (i) low temperatures;
      (ii) high temperatures?

(6) In the circuit of *Figure 8.5.9*, calculate the range of output frequencies which the circuit can produce.

(7) Design an op-amp-based Wien Bridge oscillator which will produce a 3 kHz sine wave at an amplitude of $2\,V_{RMS}$.

(8) Design an op-amp-based Wien Bridge oscillator which will produce either a 256 Hz tone or a 440 Hz tone (*Figure 8.5.10*).

(9) Design an op-amp-based Wien Bridge variable frequency oscillator which can output a range of sine waves from 300 Hz to 3400 Hz.

8.6 Relaxation oscillators

Resistor–capacitor timing

When a step voltage is applied to the resistor, the voltage builds up on the capacitor according to the exponential growth formula:

$$V_C = V_S\left(1 - e^{-t_1/RC}\right)$$

The formulae for each of the times to reach voltages $V1$ and $V2$ is given by:

$$V_1 = V_S\left(1 - e^{-t_1/RC}\right) \qquad V_2 = V_S\left(1 - e^{-t_2/RC}\right)$$

$$e^{-t_1/RC} = 1 - \frac{V_1}{V_S} \qquad e^{-t_2/RC} = 1 - \frac{V_2}{V_S}$$

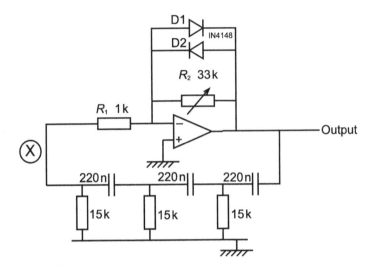

Figure 8.5.5

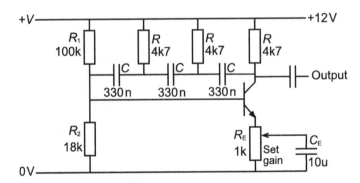

Figure 8.5.6

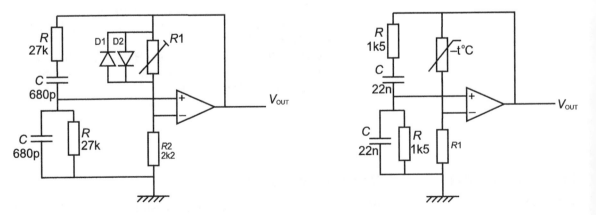

Figure 8.5.7   **Figure 8.5.8**

Oscillators

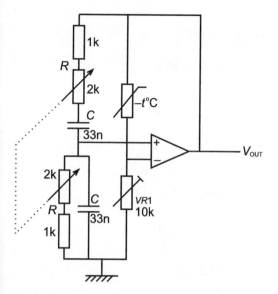

Figure 8.5.9

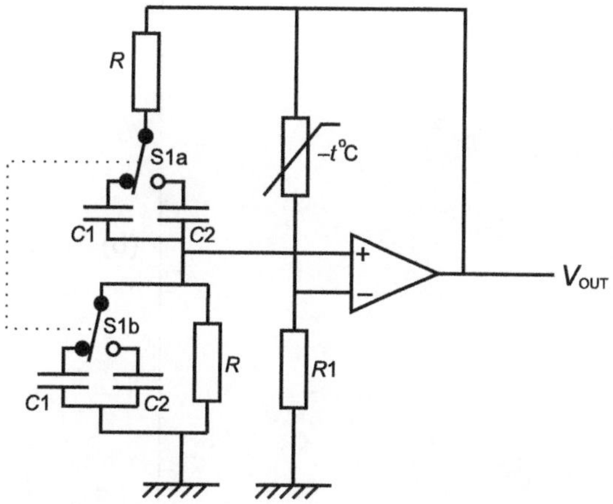

Figure 8.5.10

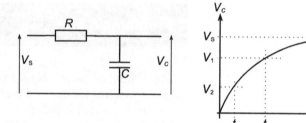

Figure 8.6.1 *An RC timing circuit*

Dividing these equations:

$$\frac{e^{-t_1/RC}}{e^{-t_2/RC}} = \frac{\dfrac{V_S - V_1}{V_S}}{\dfrac{V_S - V_2}{V_S}}$$

$$e^{(-t_1/RC)-(-t_2/RC)} = \frac{V_S - V_1}{V_S - V_2}$$

$$e^{t_2 - t_1/RC} = \frac{V_S - V_1}{V_S - V_2}$$

$$\frac{t_2 - t_1}{RC} = \ln\left(\frac{V_S - V_1}{V_S - V_2}\right)$$

$$t_2 - t_1 = RC \ln\left(\frac{V_S - V_1}{V_S - V_2}\right)$$

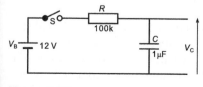

Figure 8.6.2

Example 8.6.1

In the simple *RC* network of Figure 8.6.2, calculate:

(a) how long after 12 V is applied to the input resistor the voltage across the capacitor will rise to 3 V;

(b) how long it will then take to rise to 9 V.

(a)
$$V_C = V_B\left(1 - e^{-t/RC}\right)$$
$$t = -RC \ln\left(\frac{V_B - V_C}{V_B}\right)$$
$$= -10^{-6} \times 10^5 \times \ln\left(\frac{12 - 3}{12}\right)$$
$$= 29\,\text{ms}$$

(b)
$$t_2 - t_1 = RC \ln\left(\frac{V_B - V_1}{V_B - V_2}\right)$$
$$= 0.1 \times \ln\left(\frac{9}{3}\right)$$
$$= 110\,\text{ms}$$

Problems 8.6.1

(1) Calculate how long it takes for the voltage across the capacitor to rise to 4 V after a 9 V supply is connected to the input resistor.

(2) Calculate how long it takes for the capacitor voltage to rise from 5 V to 10 V when 15 V is applied to the RC network.

(3) A 47 µF capacitor is charged to +30 V. Calculate how long it will take for its terminal voltage to fall to:
  (a) 15 V from the 30 V initial charge;
  (b) 10 V from this 15 V.

(4) Design an RC network which will take 5 s to charge a capacitor to 5 V from a 9 V supply.

The operational amplifier astable multivibrator

Astable means 'not stable', i.e. not at one fixed voltage level. A multivibrator switches between two or more states.

In this context, the circuit of Figure 8.6.3 is an oscillator which has two output levels, the positive and negative saturation output voltages of the operational amplifier. It is sometimes called a relaxation oscillator.

Assume that at start-up ($t = 0$) the output is negative. It will stay negative until the voltage at B < A. Capacitor C discharges through R1 until B < A, then the output switches to its positive saturation value. R1 now charges capacitor C and the output will switch negative when voltage at B > A (Figure 8.6.4).

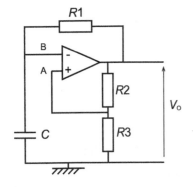

Figure 8.6.3 *The astable multivibrator*

Figure 8.6.4

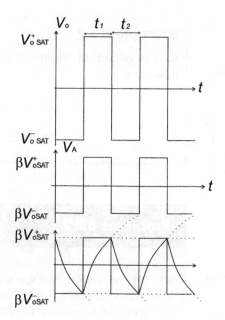

The voltage at A is given by:

$$V_A = V_O \frac{R_3}{R_2 + R_3}$$

If the feedback ratio is called β in the usual way:

$$\beta = \frac{R_3}{R_2 + R_3}$$

The time to charge a capacitor from a supply of V_S volts from an initial voltage $V1$ to a final voltage $V2$ is given by:

$$t = RC \ln\left(\frac{V_S - V_1}{V_S - V_2}\right)$$

So:

$$t_1 = R_1 C \ln\left(\frac{V_{O\,SAT}^+ - \beta V_{O\,SAT}^-}{V_{O\,SAT}^+ - \beta V_{O\,SAT}^+}\right)$$

In most amplifiers, $V_{O\,SAT}^+ = -V_{O\,SAT}^-$, so:

$$t_1 = R_1 C \ln\left(\frac{1 - -\beta}{1 - +\beta}\right) = R_1 C \ln\left(\frac{1 + \beta}{1 - \beta}\right)$$

This will be the same as t_2:

$$t_1 = R_1 C \ln \frac{1 + \dfrac{R_3}{R_2 + R_3}}{1 - \dfrac{R_3}{R_2 + R_3}} = R_1 C \ln \frac{R_2 + R_3 + R_3}{R_2 - R_3 + R_3}$$

$$= R_1 C \ln\left(1 + 2\frac{R_3}{R_2}\right)$$

The period of oscillation is $t_1 + t_2$ and is given by:

$$T = t_1 + t_2 = 2R_1 C \ln\left(1 + 2\frac{R_3}{R_2}\right)$$

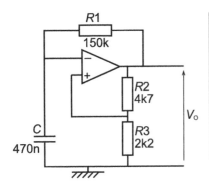

Figure 8.6.5

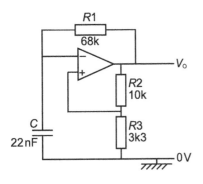

Figure 8.6.6

Example 8.6.2

In the circuit of Figure 8.6.5, calculate the frequency of operation.

$$f = 2R_1 C \ln\left(1 + 2 \times \frac{2k2}{4k7}\right)$$

$$= 10.7\,\text{Hz}$$

Problems 8.6.2

(1) In the circuit of Figure 8.6.6, $R1 = 68k$; $C1 = 22\,\text{nF}$; $R2 = 10k$; $R3 = 3k3$. Calculate:
 (a) the frequency of oscillation;
 (b) the amplitude of the square wave at the $-V_{in}$ input of the op amp.
(2) Design a square wave relaxation oscillator which produces a 10 kHz square wave output.
(3) If the op amp of question 2 is powered from a $+12\,\text{V}$ and $-12\,\text{V}$ supply, show how to modify the output to produce a 0 to $+5\,\text{V}$ square wave.
(4) Design a 1 Hz square wave oscillator. Show how to modify the output to drive two alternately flashing LEDs.

The discrete transistor astable multivibrator

In the circuit of Figure 8.6.7, the transistors operate in their saturated (or switching) mode. There are two stable states of the transistors in the oscillator:

(i) TR1 ON   TR2 OFF
(ii) TR1 OFF   TR2 ON

Figure 8.6.8 shows how the base and collector voltages of transistor Q2 vary as the switching proceeds. Q2's conduction is controlled by its base current which is related to the base voltage. This is the node

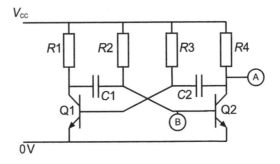

Figure 8.6.7 *The astable multibibrator*

Figure 8.6.8 The output switching states of points A and B in the astable circuit

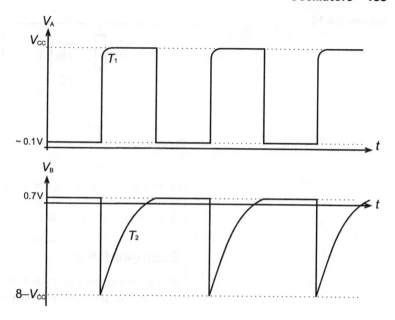

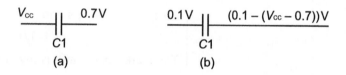

Figure 8.6.9 Q2 base voltage

voltage between $C1$ and $R2$. Figure 8.6.9 shows how the Q2 base voltage is controlled by the voltage across $C1$.

When Q1 is OFF, $C1$ is charged as shown in Figure 8.6.9(a): Q1 is OFF and Q2 is ON. Its base–emitter voltage will rise to around 0.7 V as it is in saturation. If for some reason (discussed shortly), Q1 turns ON, then at the instant of turn-on, the voltage across $C1$ becomes as shown in Figure 8.6.9(b).

Point B would then be at a **negative** voltage as shown in Figure 8.6.8. Q2 stays OFF as $C1$ charges through $R2$. When point B reaches 0.7 V, Q2 turns ON. This forces the base of Q1 negative, turning Q1 OFF.

The time to charge to 0.7 V is given by:

$$t = R_2 C_1 \ln\left(\frac{V_S - V_1}{V_S - V_2}\right)$$

$$= R_2 C_1 \ln\left(\frac{V_{CC} - (0.8 - V_{CC})}{V_{CC} - 0.7}\right)$$

$$= R_2 C_1 \ln\left(\frac{2V_{CC} - 0.8}{V_{CC} - 0.7}\right) \text{s}$$

Provided that $V_{CC} > 0.8$:

$$t = R_2 C_1 \ln 2$$
$$= 0.69\, R_2 C_1$$

If $R_2 = R_3$ and $C_1 = C_2$, then the period of oscillation is given by $1.38 RC$:

Figure 8.6.10

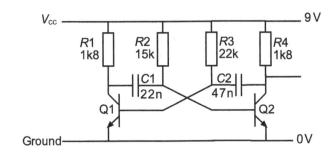

(1) For a 50% mark–space ratio, make $R2 = R3$; $C1 = C2$.
(2) Mark–space ratio = 10 % MIN and 90 % MAX.
(3) $R_2(R_3) > R_1(R_4)$ otherwise the circuit will not oscillate.

Example 8.6.3

In the circuit of Figure 8.6.10, calculate the frequency of the output and its mark–space ratio.

$$t_1 = 0.69 R2\,C1 \qquad t_2 = 0.69 R3\,C2$$
$$= 228\,\mu s \qquad\qquad = 713\,\mu s$$

So $t_1 + t_2 = 941\,\mu s$ and:

$$f = 1/t = 1.063\,\text{kHz}$$

The mark–space ratio is given by:

$$\text{Ratio} = \frac{713\,\mu s}{914\,\mu s} = 76\%$$

Problems 8.6.3

(1) In the circuit of Figure 8.6.11, calculate the frequency of the output and its mark–space ratio.
(2) Design a 1 kHz oscillator with 50:50 mark–space ratio to operate from 5 V.
(3) Design an oscillator which drives two LEDs (on the collector circuits) which flash at a 1 Hz rate with a 333 ms : 667 ms ratio.

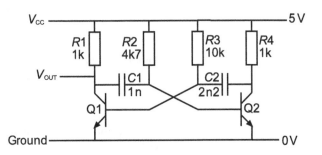

Figure 8.6.11

Figure 8.6.12 *A transistor monostable multivibrator*

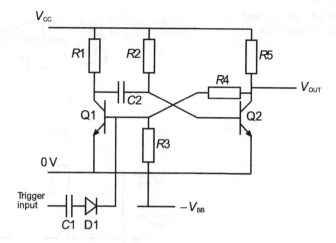

Monostable multivibrators

These devices form a special case of the astable variety. A monostable multivibrator will change state once only and then return to its original state (hence 'mono'). It is a single pulse generator (what used to be called a 'one-shot').

In Figure 8.6.12, the 'normal' situation is that $R2$ holds Q2 ON. This means that there is not enough base drive for Q1 which is held OFF. To ensure this, an additional negative supply is often connected to Q1. A short positive pulse into the base of Q1 turns it ON. As with the astable circuit, the collector of Q1 drops nearly to zero which drives the right-hand side of $C2$ negative, as shown in Figure 8.6.13. $C2$ eventually recharges through $R2$ and only when the voltage at the $R2C2$ junction passes the turn-on voltage of Q2 (0.6 to 0.7 V) will the circuit switch back to its stable state. The mathematics is exactly the same as for the astable circuit and the same reasoning process will result in a pulse period of:

$$t = 0.69 R_2 C_2$$

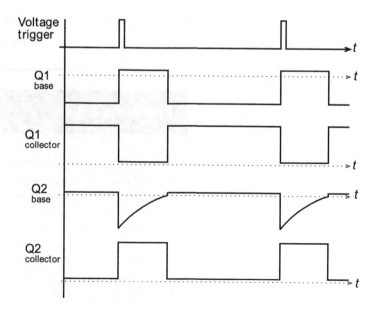

Figure 8.6.13

Figure 8.6.14 *An op amp monostable multivibrator*

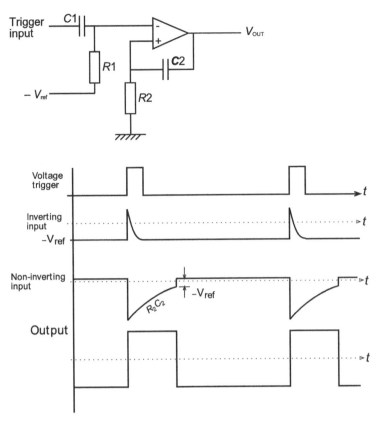

Figure 8.6.15

The circuit could also be built with an operational amplifier as shown in Figure 8.6.14.

Example 8.6.4

Calculate the pulse width of the output waveforms in Figure 8.6.4 following a pulse applied to the trigger input. In this circuit, $C = 10\,nF$; $R = 10\,k$.

The pulse width is given by the relationship:

$$t = 0.69 \times R_2 C_2$$
$$= 69\,\mu s$$

Problems 8.6.4

(1) If $C2 = 15\,nF$ and $R2 = 33\,k$, calculate the pulse width of the output of Figure 8.6.14.
(2) Design an operational-amplifier-based monostable to provide an output.
(3) Calculate the pulse width of the output of the circuit when the input is triggered.
(4) Design a two-transistor monostable which operates from a 12 V supply and can be triggered to produce a 1 ms output pulse.

Figure 8.6.16

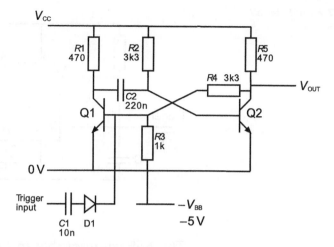

8.7 The 555 timer

Any book with a section on timers in the form of monostable or astable multivibrators will usually include a section on this versatile integrated circuit. This venerable device has been in existence for over two decades and has alone been the subject of many books. It is basically a device which uses a resistor and capacitor to determine the timing period. Figure 8.7.1 shows the pinout and internal schematic.

The input stages consist of three equal value resistors connected to two comparators. Comparator 1 detects when the threshold voltage is at two-thirds of V_{CC}, while comparator 2 detects when the trigger voltage is at one-third of V_{CC}. The control voltage is normally connected to a smoothing capacitor, but can be used in some applications to vary the reference voltage into comparator 1. The reset resets the circuit. If the reset input is at V_{CC}, the circuit will work, but it will not if the reset is at 0 V. The output goes high when the latch is SET, and low when the latch is RESET. The discharge output transistor (internal to the 555) is turned ON when the latch is RESET and is off when the latch is SET. It is normally used to discharge the external timing capacitor. Figure 8.7.2 shows how this is applied in a 555-based monostable multivibrator.

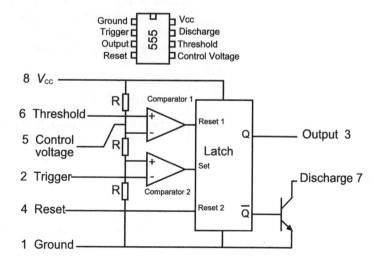

Figure 8.7.1 *The 555 timer*

Figure 8.7.2 *555-based monostable*

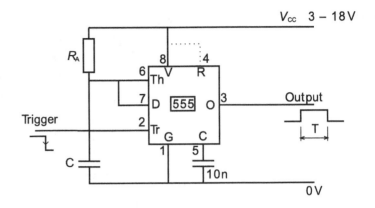

The 555 monostable multivibrator

To start with, the trigger input should be high. The output will normally be low (0 V) and the internal discharge transistor is ON which shorts out the capacitor C to prevent it from charging up. When the trigger input drops to less than one-third of V_{CC}, then the monostable is triggered and starts timing. The output goes high (V_{CC}) and the discharge capacitor releases its hold on the timing capacitor. The voltage on C builds up until it reaches two-thirds of V_{CC}. This is the threshold level. At this point, the latch is reset and this causes the output to fall to 0 V, and the discharge transistor turns ON and discharges the capacitor. The timing period is given by:

$$T = 1.1 \times R_A \times C \text{ s}$$

The nature of the input circuitry means that R_A must not be less than 5k. The capacitor can be as large as is necessary to provide the required timing. Output pulses of up to an hour are possible, but the repeatability is somewhat suspect. The output capability of the 555 is that it can source or sink the relatively high current of 200 mA. This is perfectly sufficient for LEDs or relays or even small alarm sounders. One caution is that the input trigger pulse must be smaller than the designed output pulse. If it is not, then the output will simply stay high for as long as the input trigger is low. A common modification which allows trigger pulses of any length is shown in Figure 8.7.3(a). The input can be 'pulled down' with a switch and the input to the trigger will only go low as set by the 10k–10n time constant. This circuit allows several 555s to be cascaded together if required to make a sequencer.

Example 8.7.1

Design a circuit which, at the press of a button, will provide a 10 s output pulse to turn on a 12 V 100 mA relay coil.

$$T = 1.1 \times R_A \times C$$
$$10 = 1.1 \times R_A \times C$$

10 s is a relatively long delay, so make $C = 100\,\mu\text{F}$.

From this, R_A works out to 91k. Electrolytic capacitors are not manufactured to any great tolerance, so an allowance of at least 20% should be made on the resistor value – say a 75k fixed in series with a 50k variable.

Figure 8.7.4 shows the solution. Note the back emf diode across the relay coil.

Problems 8.7.1

(1) Design a monostable oscillator which provides an output pulse to light an LED for 0.5 s when the input button is pressed.
(2) Design a sequencer which provides outputs of 1 s, 3 s and 5 s. Is it possible to link the final output back into the first input to make a continuous system?
(3) Prove the relationship $T = 1.1RC$.

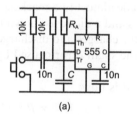

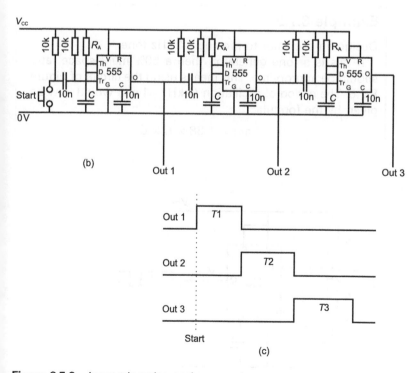

Figure 8.7.3 *Input triggering and sequencing*

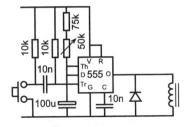

Figure 8.7.4

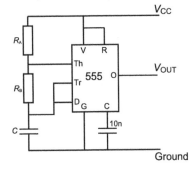

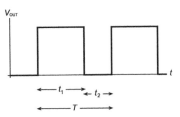

Figure 8.7.5 The 555 astable

The 555 astable multivibrator

An additional timing resistor, R_A, has been added to this circuit in Figure 8.7.5. The position of the discharge transistor is such that when the capacitor is discharged, it does so through C and R_A. When the capacitor charges up, it does so through R_A and R_B until the threshold voltage is two-thirds of the supply voltage. It then discharges the capacitor until the trigger voltage of one-third of the supply voltage is reached. The output of this astable does not show a 50% mark–space ratio.

The output voltage is high for a period:

$$t_1 = 0.69 \times (R_A + R_B) \times C$$

and the output is low for a period of:

$$t_2 = 0.69 \times R_B \times C$$

The full period of oscillation is given by:

$$T = t_1 + t_2 = 0.69 \times (R_A + 2R_B) \times C$$

The manufacturers place a practical minimum limit on R_B of 3 kΩ.

The output can be made almost to a mark–space ratio of 50% by using two equal resistors and a diode as shown in Figure 8.7.6. This ensures that the capacitor charges and discharges through the same value of resistor (with the slight error caused by the diode forward voltage drop).

In this case, the period is now:

$$T = t_1 + t_2 = 1.38 \times R \times C$$

Example 8.7.2

Design an oscillator to provide a 1 kHz tone.

If this is a tone generator, then a 50% mark–space ratio should be appropriate. This means using the circuit of Figure 8.7.6. The period of oscillation (1 kHz = 1 ms) would now be given by the formula:

$$1\,\text{mS} = 1.38 \times R \times C$$

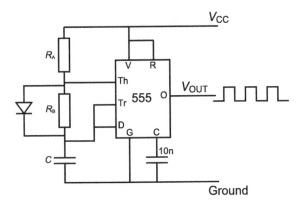

Figure 8.7.6

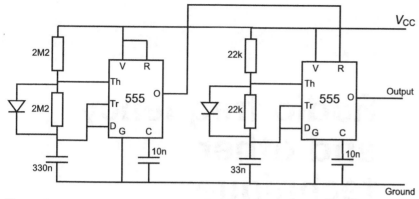

Figure 8.7.7

If we choose C to be 33 nF, then R will be 22 kΩ.

Example 8.7.3

Design an oscillator to provide a 1 kHz alarm tone pulsed on and off every 1 s.

This could be solved by using the output of a 1 s oscillator to drive the reset line of a second 555 which had been designed to oscillate at 1 kHz. A possible solution is shown in Figure 8.7.7.

Problems 8.7.2

(1) Design an oscillator to generate a 3 kHz square wave.
(2) Design a variable oscillator which will generate a frequency in the range of 300–3400 Hz.
(3) Use a comparator and a 555 to generate a 4 kHz square wave whenever a signal input voltage goes lower than 4.7 V.
(4) Design a gated oscillator with a monostable and an astable which will produce a 750 Hz tone for 10 s after a button is pressed. What modification will stop the tone at the press of a second button? (No, not the power switch!)
(5) Why is the formula for the time period of the 555 independent of the supply voltage?
(6) Design a sequencer which outputs four 0.5 s rising tones of 262 Hz, 330 Hz, 392 Hz and 524 Hz when the start button is pressed.

9 Radio frequency and other techniques

Summary

Radio frequency can mean almost any frequency from kHz up to hundreds of GHz. As far as this chapter is concerned, we will look at those techniques which are necessary above 1 MHz. This is where the inter-electrode capacitance of transistors starts to become significant.

9.1 Transistor circuits

Technically, the r.f. world is subdivided as shown in Figure 9.1.1. The letter designations for r.f. bands are divided as shown in Table 9.1.1.

Table 9.1.1

Frequency band	Frequency
A	100–300 MHz
B	300–500 MHz
C	0.5–1 GHz
D	1–2 GHz
E	2–3 GHz
F	3–4 GHz
G	4–6 GHz
H	6–8 GHz
I	8–10 GHz
J	10–20 GHz
K	20–40 GHz
L	40–60 GHz
M	60–100 GHz

kHz				MHz						GHz		
10	30	100	300	1	3	10	30	100	300	1	3	10
km				m						mm		
30	10	3	1	300	100	30	10	3	1	300	100	30
VLF	LF			MF		HF		VHF		UHF		SHF

Figure 9.1.1 *Frequency bands*

Figure 9.1.2 *The hybrid-π equivalent circuit*

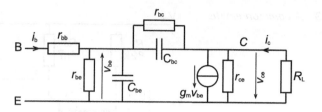

Miller effect

This is a transistor effect whereby internal feedback between electrodes degrades the performance of the amplifier containing the transistor. Figure 9.1.2 shows the 'hybrid-π' equivalent circuit of a transistor.

The components which are not always shown in simple equivalent circuits used for calculation at audio frequencies are:

r_{bb} the ohmic resistance of the base region;
r_{be} the resistance of the forward biased base–emitter junction;
C_{be} the capacitance of the forward biased base–emitter junction;
C_{bc} the capacitance of the reverse biased collector–base junction;
r_{bc} the resistance of the reverse biased collector–base junction;
r_{ce} the resistance of the output when the base–emitter terminals are short circuited.

For a 2N3904 transistor, the capacitance figures might typically be:

$$C_{bc} = 4\,\text{pF} \quad C_{be} = 8\,\text{pF}$$

The resistance figures might typically be:

$$r_{bb} = 300\,\Omega \quad r_{be} = 2000\,\Omega \quad r_{bc} = 1.5\,\text{M}\Omega \quad r_{ce} = 25\,\text{k}\Omega$$

All of these values are dependent on many factors, including temperature, bias current and bias voltage.

At low frequencies, the capacitance values have a very large impedance. 8 pF is 20 MΩ at 1 kHz. As the frequency rises, these impedances fall, and there is a lower impedance path from the base to the emitter, which shunts out the input signal and from the collector to the base, which is a negative (gain reducing) feedback path. The current flowing through C_{bc} will be given by:

$$V = i \times X_C$$

$$v_{bc} - v_{ce} = i \times \frac{1}{2\pi f C_{bc}}$$

$$i = 2\pi f C_{bc}(v_{bc} - v_{ce})$$

Now

$$v_{ce} = -g_m R_L v_{bc}$$

so

$$i = 2\pi f C_{bc}(v_{bc} + v_{bc} g_m R_L)$$

or

$$i = 2\pi f C_{bc} v_{bc}(1 + g_m R_L)$$

The admittance looking into the circuit at v_{be} is given by:

$$Y = \frac{i}{v_{bc}} = 2\pi f C_{bc}(1 + g_m R_L)$$

Figure 9.1.3 *A common emitter amplifier*

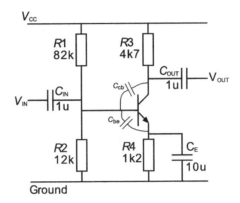

This apparent magnification of the base–emitter capacitance by the $(1 + g_m R_L)$ term is called the **Miller effect**. In practice, it causes the frequency response of the amplifier to roll-off at high frequencies. The circuit of Figure 9.1.3 shows a common emitter amplifier. The internal capacitance C_{be} has the same effect as a low pass filter, while the internal capacitance C_{ce} provides negative feedback at high frequencies. When a real capacitor is connected in this position deliberately, it is to limit the high frequency response of the amplifier.

The common base amplifier

To improve the high frequency performance, the effect of C_{bc} must be reduced, and this can be usefully done with the common base amplifier as shown in Figure 9.1.4.

In this circuit, the base is held at a constant potential and the signal input at the emitter. The transistor amplifies the changes of current caused by the variation of base–emitter voltage, so in effect, it does not matter much whether the emitter is held constant and the voltage on the base varied, or whether the base is held constant and the voltage on the emitter varied. There will still be a voltage gain, but the frequency range over which the signal can be amplified is greatly increased. The downside is that this circuit has a poor current gain. The input seen in Figure 9.1.4 feeds a tuned circuit into the low input impedance common base stage. The collector output also has a tuned stage to improve the gain and selectivity of the circuit. The gain can be improved without compromising the noise performance with the cascode connection (Figure 9.1.5).

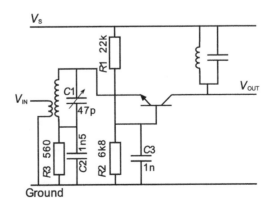

Figure 9.1.4 *The common base amplifier*

Figure 9.1.5 *The BJT cascode*

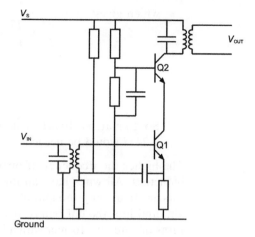

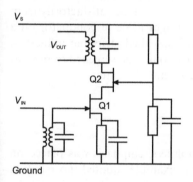

Figure 9.1.6 *The FET cascode*

The two transistors are effectively in series across the supply, with the common emitter stage of Q1 feeding into the low input impedance of the common base stage of Q2. The low input impedance of the common base stage is such that the gain of the first stage is low, so the Miller effect is reduced. A cascode amplifier made with FETs usually has a better noise performance than the circuit constructed with BJTs (Figure 9.1.6).

R.f. transformers

In all of these circuits, note the proliferation of transformers. These are used in r.f. work to:

(i) **Provide interstage coupling**
A.c. signals have to be coupled from one part of the circuit another without allowing the DC bias levels of the one circuit from interfering with the next stage. At audio frequencies, capacitors provide a cheap and light method, and transformers would be bulky. As the frequency increases, the opposite becomes true.

(ii) **Provide gain and selectivity through resonant circuits**
One or both sides of the transformer can be used to resonate with a parallel capacitor. The resonant frequency as discussed in Chapter 8 produces a circuit which has a narrow bandwidth and a high dynamic impedance. The gain of the transistor stage is proportional to the value of the output load, so this high dynamic impedance increases the gain.

(iii) **Provide impedance matching between stages**
One important use of a transformer is its ability to have a different impedance at each set of windings (Figure 9.1.7). In the transformer shown, the turns ratio is $N:1$, so:

$$\frac{V1}{V2} = \frac{N}{1} \quad \text{and} \quad \frac{I1}{I2} = \frac{1}{N}$$

Dividing these:

$$\frac{V1/V2}{I1/I2} = \frac{N/1}{1/N}$$

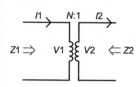

Figure 9.1.7

which gives:

$$\frac{V1}{I2} \times \frac{I2}{V2} = N^2$$

$$Z1 \times \frac{1}{Z2} = N^2$$

$$Z1 = Z2 \times N^2$$

So, to match a 1000 Ω load to a 200 Ω source, a turns ratio of 2.24 : 1 would be needed.

The higher the frequency of operation, the smaller and lighter the inductors. All wires have an inherent inductance, although inductors are frequently thought of as coils of wire. At the higher end of vhf, and into the uhf bands, the inductors are no more than a few turns of wire on a former with a ferrite core which can be screwed in or out of the core to provide an element of adjustability. As the frequencies get higher, the inductor can be made satisfactorily from a straight piece of wire or even a straight piece of conductor on a printed circuit board. These are the so-called strip-line inductors. Circuits containing these usually have to be well screened to prevent stray fields from coupling with the circuit elements.

9.2 Operational amplifiers

In the chapters on operational amplifiers, emphasis was placed on the gain–bandwidth product. In the examples quoted, no mention was made of the high frequency capability of certain special purpose operational amplifiers. There are video amplifier ICs which operate with a fixed gain and there are also true operational amplifiers which provide a very high gain–bandwidth product. In between the discrete and monolithic amplifiers are a range of hybrid amplifiers composed of discrete elements integrated into one device on a ceramic substrate. These offer good performance, but at a much higher price than a fully monolithic integrated circuit (monolithic = on one piece of silicon).

For example, the Philips Semiconductor NE5539 is a monolithic uhf operational amplifier (Figure 9.2.1). It features a 1.2 GHz GBW at an open loop gain of 52 dB and a slew rate of 600 V/μs. As is common with all devices of this sort, the layout on a circuit board is extremely critical. The manufacturers recommend using a good specification double-sided printed circuit board which will provide a ground plane.

More recently, the 8 pin Maxim MAX477 provides a full power bandwidth of 200 MHz and a slew rate of 1100 V/μs (Figure 9.2.2). Once again, with this device, layout is all-important. The manufacturer recommends:

- Do not use wire-wrap boards. They are too inductive.
- Do not use IC sockets. They increase parasitic capacitance and inductance.
- In general, surface-mount components have shorter leads and lower parasitic reactance, giving better high-frequency performance than through-hole components.
- The PC board should have at least two layers, with one side a signal layer and the other a ground plane.

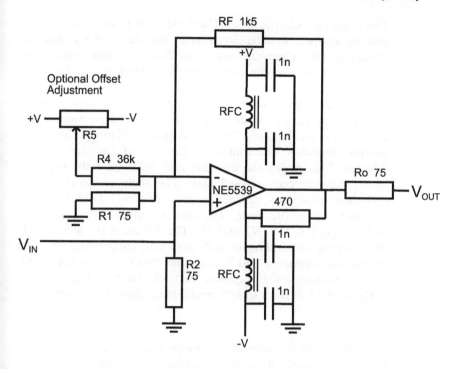

Figure 9.2.1 *Manufacturers' recommended layout of the NE5539*

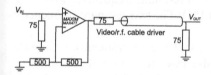

Figure 9.2.2 *The MAX477*

- Keep signal lines as short and straight as possible.
- Do not make 90° turns; round all corners.
- The ground plane should be as free from voids as possible.

Surface-mount resistors are best for high-frequency circuits. Their material is similar to that of metal-film resistors, but to minimize inductance it is deposited in a flat, linear manner using a thick film. Their small size and lack of leads also minimize parasitic inductance and capacitance.

Driving large capacitive loads increases the chance of oscillations in most amplifier circuits. This is especially true for circuits with high loop gain, such as voltage followers. A second problem when driving capacitive loads results from the amplifier's output impedance, which looks inductive at high frequency. This inductance forms an *LC* resonant circuit with the capacitive load, which causes a peak in the frequency response. To drive larger-capacitance loads or to reduce ringing, add an isolation resistor between the amplifier's output and the load.

9.3 Crystal oscillators

Crystal oscillators are included in this section purely because they tend to be high frequency references used for microprocessors and radio work. The piezoelectric effect in quartz was discovered by Pierre Curie as long ago as 1880, but it was not used until the 1920s to control the frequency of an oscillator.

A sliver of quartz will mechanically deform when a voltage is applied to opposite faces. Alternatively, if a mechanical force is applied to opposing faces, a voltage is generated across the crystal.

This is the piezoelectric effect. The device thus formed can be made into an acoustic sounder (such as in the alarm of watches and clocks) or into a pressure sensor mechanism. These applications aside, the most common use of the piezo effect is as a frequency reference.

The sliver of quartz is cut from a bar of quartz at precise angles to the crystal axis and polished prior to the vacuum deposition of gold electrodes to the flat faces. The disc of quartz will now resonate at a natural frequency determined by the size of the slice and its thinness. The smaller the diameter and the thinner the slice, the higher the frequency of oscillation. The angle of the cut to the crystallographic axis determines the use and stability of the resultant crystal oscillator. The cuts are given letter references such as AT, CT, DT, NT and SC. The AT cut is the most widely used because the resulting crystals show little temperature variation of resonant frequency and because the crystals can be cut in this way to resonate in the frequency range of 1 to 200 MHz.

Figure 9.3.1 shows a quartz crystal equivalent circuit, where:

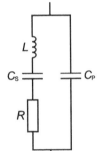

Figure 9.3.1 *Quartz crystal equivalent circuit*

L represents the dynamic or motional inductance;
C_S represents the dynamic or motional capacitance;
R is the equivalent series resistance;
C_P is the parallel or static capacitance.

For a 2 MHz crystal, $L = 520\,\text{mH}$, $C_S = 12\,\text{fF}$, $R = 100$ and $C_P = 4\,\text{pF}$.

The resultant component has two resonant frequencies which are closely spaced (usually to within 1%). The two resonant frequencies are shown in Figure 9.3.2.

The reactances are either zero for the series resonance or infinite for the parallel resonance, and this leads to a state of minimum or maximum impedance for the crystal equivalent circuit.

The mechanical stability ensures that the finished slice of quartz has a high Q factor and excellent stability. Without any special protection, a stability of 5–20 ppm (parts per million) can be achieved. In this case, it represents 0.005–0.02% stability. To do better than this, temperature compensated devices and techniques are required which improve the stability to 0.2–5 ppm. The very best crystal oscillators are oven controlled and can achieve 0.001–1 ppm

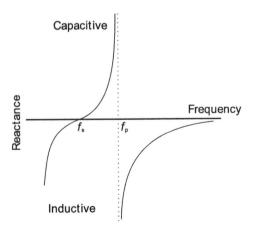

Figure 9.3.2 *Quartz crystal resonant frequencies*

Figure 9.3.3 *Crystal packages*

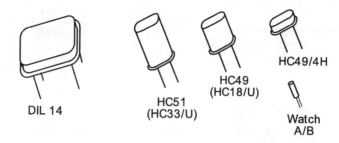

stability, though at a price – they are very expensive, have a high power consumption and take a long time to warm up.

The most common crystal in commercial production is probably resonating at 32 768 Hz. When this is divided by 2^{15}, the result is one pulse per second, which is ideal for wrist watches, clocks and other time-keeping devices. The frequency can be 'pulled' slightly by connecting a series or parallel capacitance.

There are lots of different frequencies of oscillation of crystals available off-the-shelf. If specific frequencies are required it becomes a slightly more expensive order, since the crystal slice has to be individually lapped to produce the correct output. Figure 9.3.3 shows the common crystal packages.

The most stable (non-temperature-regulated) crystals tend to be vacuum sealed into a glass envelope. At the lower end of the scale is a simple solder seal. This still performs well, but contamination of the crystal can result in ageing. Between the two comes the welded cans – resistance welded, or, for better performance, cold welded.

Oscillators

The model shown in Figure 9.3.1 has two resonance frequencies – series and parallel – which produce the lowest and highest impedances respectively. Figure 9.3.4 shows a form of Pierce oscillator. The r.f. choke provides a load resistance at radio frequencies. The crystal provides signal feedback at its resonant frequency.

A FET can be the basis of a crystal oscillator as shown in Figure 9.3.5. This is a variant of the Clapp oscillator which is itself developed from the Colpitts oscillator. Here, the LC resonant

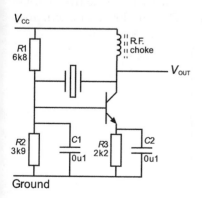

Figure 9.3.4 *A Pierce oscillator*

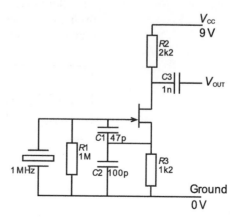

Figure 9.3.5 *The Clapp oscillator*

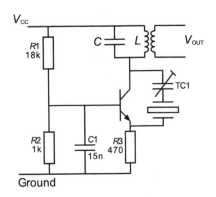

Figure 9.3.6 *The Butler oscillator*

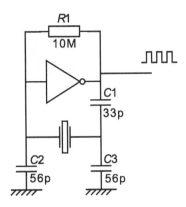

Figure 9.3.7 *The CMOS oscillator*

circuit is replaced by a crystal and a proportion of the drive is applied between the gate and the source.

Sometimes it is advantageous to operate on a harmonic of the resonant frequency. Figure 9.3.6 is called a Butler overtone circuit. The LC collector load is tuned to the harmonic of the resonant frequency – usually the third or fifth overtone is selected. The trimmer allows for a slight pulling of the crystal frequency.

In Chapter 10 we will meet the logic circuits which switch from one state to another – usually between 0 and 5 V. In digital circuits which require a square wave clock to drive timing or sequential circuits, a crystal derived clock is often used. Two circuits follow which are suitable for the two main logic families – those based on transistor–transistor logic (TTL) and those based on complementary MOSFETs (CMOS). Figure 9.3.7 is a type of Colpitts oscillator where the crystal has replaced the inductor, and the CMOS inverter has been linearized to perform like an inverting amplifier by the 10 MΩ resistor. The π network formed by the crystal and $C_2//C_3$ provide the other 180° phase shift to provide positive feedback.

The same technique is used for the TTL inverter, but the components are modified slightly to cope with the different characteristics of the TTL inverter gate (Figure 9.3.8).

Ceramic resonators

An alternative to crystals is a range of devices called ceramic resonators. These piezoelectric components are made from barium titanate or lead-zirconium titanate. They are low-cost alternatives

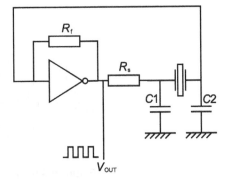

Figure 9.3.8 *The TTL oscillator*

Figure 9.3.9 *The CMOS oscillator for a ceramic resonator*

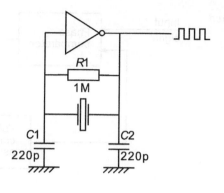

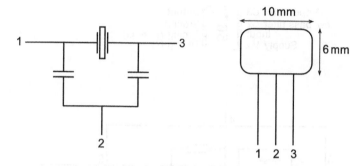

Figure 9.3.10

to quartz, but only have a frequency tolerance of ±0.5% compared with the quartz crystal's 0.002%. A 4 MHz ceramic resonator would have an equivalent circuit component count of:

$$R = 8\,\Omega \quad L = 318\,\text{mH} \quad C_S = 5.4\,\text{pF} \quad C_P = 42\,\text{pF}$$

The Q factor is a factor of ten lower than its quartz equivalent at 1000. Oscillators are easily constructed as in Figure 9.3.9, which is set up for the CMOS inverter.

There is a range of ceramic resonators which have $C1$ and $C2$ built in. Each one has three pins and is connected as in Figure 9.3.10. They are convenient and cheap, but are only available in a restricted frequency range (1–20 MHz, dependent on manufacturer). Two-terminal resonators are available in the slightly larger range of 180 kHz–20 MHz.

9.4 Phase locked loops

A phase locked loop is an electronic control loop as shown in Figure 9.4.1. It contains a phase comparator, a low pass filter and a voltage controlled oscillator. These could be separate circuit blocks, although the neatest solution is obtained by integrating them onto one piece of silicon as in the Philips Bipolar ICs NE567 or NE565 or the CMOS HEF4046. (Actually, not all of the phase locked loop circuit is integrated onto the silicon: the low pass filter is usually kept separate to help with the flexibility of the design.) The circuit is provided with an a.c. signal and the loop drives its internal voltage controlled oscillator until it is oscillating at the same frequency as the input.

It would appear that this circuit has one input but no output. In Figure 9.4.1 that is the case, but the output could be derived from the output of the low pass filter or the voltage controlled oscillator.

Figure 9.4.1 *The phase locked loop*

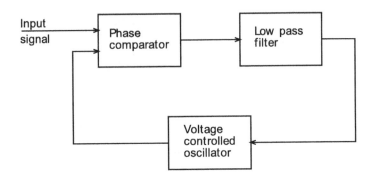

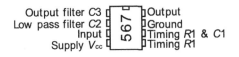

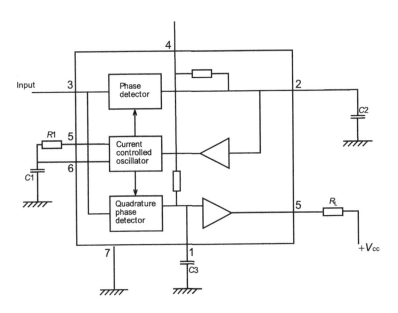

Figure 9.4.2 *The 567 PLL*

It all depends on the circuit function, and PLLs have a wide range of applications.

Some common uses

- **Tone decoder** This detects tones i.e. frequencies within a specified range. The voltage controlled oscillator stage is usually limited in the range of frequencies over which it can act. Thus, it can detect whether or not a tone is present.
- **FM demodulator** In frequency modulation, the low frequency information signal changes (modulates) the frequency of a higher frequency **carrier**. For example, if a 100 kHz carrier was frequency modulated by a 1 kHz tone, the resultant signal would have a frequency which varied from 99 kHz to 101 kHz. The phase lock loop can lock to the carrier and detect the changes of the information signal.

- **AM demodulator** In amplitude modulation, the low frequency information changes the amplitude of the high frequency carrier. The phase lock loop locks to the carrier frequency. The signal (carrier + information) is then multiplied by the PLL output (carrier only) and the information can be extracted from the resultant waveform. This is called a **homodyne** demodulator.
- **Frequency multiplication/synthesis** It is possible to force the voltage controlled oscillator to output a multiple of the input frequency. An example of this technique will be presented later.
- **Signal conditioning** Digital signals are often transmitted in the form of two tones. For example, a 300 bit/sec (300 baud) standard calls for a logic ONE to be transmitted as a 1270 Hz tone and a logic ZERO as a 1070 Hz tone. This requirement derives from the fact that telephone lines, down which these codes are transmitted, are a.c. coupled and have filters to prevent any frequency lower than 300 Hz passing. The phase lock loop is ideal for detecting when one or other of these frequencies is present. It is a development of the tone decoder.

The component parts of a phase locked loop (PLL)

- **The phase comparator** This compares the phases of the signal input and the output of the voltage controlled oscillator and provides an output which is proportional to the difference in phase between them. In a digital circuit, the output is usually a series of pulses the width of which is smoothed to an analogue voltage by the low pass filter. In an analogue circuit, the phase comparator directly outputs a voltage proportional to the phase difference. However, when out of lock, the two input frequencies to the phase comparator are 'slipping' past each other, so the analogue output of this type of phase comparator still needs a:
- **Low pass filter** This could be as simple as a first order RC network or as complicated as a sharp-cut-off IC switched capacitor filter. The analogue output of the low pass filter is fed into the:
- **Voltage controlled oscillator (VCO)** As its name suggests, this circuit element provides a voltage output which is proportional to its analogue voltage input. Some PLLs have a current controlled oscillator instead of a VCO, but apart from minor circuit changes, the function is more or less the same. If the output of the voltage controlled oscillator is not the same as the signal input in frequency and phase, then an error signal is generated to drive the VCO output in the correct direction. Even when the two frequencies are the same and the loop is said to be locked, then there is usually a small phase difference between the two inputs to the phase comparator.

The parameters of a phase locked loop

- **Free running frequency f_O** This is the frequency at which the oscillator will produce an output even when no input is present. It is determined by the external components of the voltage controlled oscillator.

- **Capture range** f_C This is the range of frequencies (usually either side of the free running frequency) for which the loop can 'lock'. A loop is in lock when it has managed to adjust the frequency of the VCO to the same as the input frequency.
- **Lock range** f_L This is the range of frequencies over which the loop can stay in lock. The lock range is actually larger than the capture range.

To be able to attempt a mathematical analysis, you would also need the parameters:

- **Phase detector gain factor** K_D The transfer function of the phase detector.
- **Natural frequency** w_n The characteristic frequency of the loop. Sometimes defined as the frequency at which the output will swing while trying to follow a step change in input.
- **VCO conversion gain** K_o The factor which describes the output frequency as a function of the input.
- **Loop gain** K_v The product of all the of the d.c. gains of the loop elements.

With no input signal, the PLL will free-run at its centre frequency. When an input is applied, the phase comparator compares the two signals and the output is filtered before being input to the VCO. Designing the low pass filter can be a compromise between the speed of lock (sometimes called the speed of capture) and the stability of the lock (i.e. its susceptibility to transient changes in the input frequency). In general, the longer the time constant of the low pass filter:

- the longer it takes to lock to an input signal;
- the capture range is reduced;
- the PLL becomes more immune to momentary fading or loss of signal;
- the PLL is better at rejecting noise and out-of-band signals.

The remainder of this chapter will concern itself with interpreting the manufacturers' applications rather than with a strict mathematical analysis.

The designer has to use these data sheets in order to set up a phase lock loop with the necessary characteristics. To start with, we shall look at the 567 PLL Tone Decoder. This is an 8-pin circuit which can cope with frequencies in the range of 0.01 Hz–500 kHz.

The centre frequency is calculated from:

$$f_O = \frac{1}{1.1 \times R1 \times C1} \quad (2\,k\Omega \leq R1 \leq 20\,k\Omega)$$

The capture bandwidth is given by:

$$f_C = 1070 \sqrt{\frac{V_{IN}}{f_O \times C2}}$$

C3 and C2 are chosen from Figure 9.4.3.

The simplest application circuit is shown in Figure 9.4.4. This uses the frequency determining components R1 and C1 and the filter capacitors C2 and C3. The output resistor is needed as a pull-up resistor. The output current capability is in excess of

Figure 9.4.3 *Detection bandwidth*

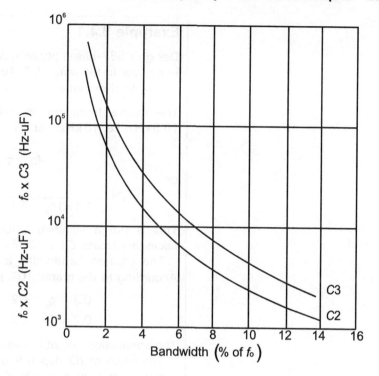

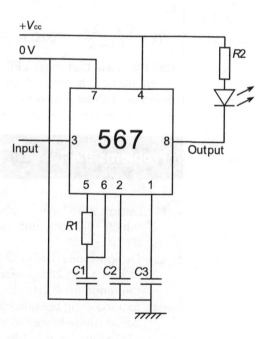

Figure 9.4.4 *The 567 tone decoder*

100 mA, and so could easily be used to drive an LED or a relay (with suitable back-emf protection!). One point to note is that the output is actually an open collector of a transistor (which is why a pull-up resistor is needed). This means that while the circuit is in lock the output is actually low (0 V). A relay or indicator placed between the $+V_{CC}$ supply and the output will turn ON when the circuit is in lock. When the circuit is not in lock, the output will be high, at V_{CC}.

Example 9.4.1

Design a 567-based phase locked loop which will lock onto a frequency in the range 9.5–10.5 kHz and drive an LED at its output to show when lock has been achieved.

The centre frequency is half-way between 9.5 kHz and 10.5 kHz, i.e. 10 kHz. Let $R1 = 10k$:

$$f_O = \frac{1}{1.1 R1 C1}$$

or

$$C1 = \frac{1}{1.1 R1 f_O} = \frac{1}{1.1 \times 10k \times 10k} = 9.09\,nF$$

This is not a standard value of capacitor, so it may be necessary to use $C1 = 10\,nF$ and $R1 = 9k1$.

The capture bandwidth is 1 kHz, which is 10% of f_O. According to the graph, 10% puts:

$$0.3 = f_O \times C2 \quad \text{or} \quad C2 = 30\,\mu F$$
$$0.7 = f_O \times C3 \quad \text{or} \quad C3 = 70\,\mu F$$

The application circuit is shown in Figure 9.4.4.

The value of $R2$ depends on the supply voltage. The 567 can operate from a supply of up to 10 V. So if a 9 V battery supply was used, then the value of $R2$ could be expected to be:

$$R2 = \frac{9V - 2V}{10\,mA} = 700\,\Omega \text{ or } 680\,\Omega \text{ (nearest preferred value)}$$

(on the basis that most LEDs will give a good light output when passing a forward current of around 10 mA and, in doing so, exhibit a forward voltage drop of around 2 V).

Problems 9.4.1

(1) Design a 567-based phase locked loop tone decoder which will lock onto a signal between 85 kHz and 87 kHz.
(2) Design a 567-based phase locked loop tone decoder to turn a 9 V, 20 mA relay ON when the frequency is in the range 49–51 Hz.
(3) Extrapolating the graphs of Figure 9.4.3, design a 567-based phase locked loop circuit which will turn on an LED when the input frequency is between 900 Hz and 1100 Hz.
(4) Design a light-beam break detector which uses a 555-based oscillator driving an LED at 7.5 kHz as the transmitter and 567-based detector with its free running frequency set to 7.5 kHz. The 567 will need a preamplifier consisting of a detector photodiode operating into a high gain amplifier. The output of the 567 should operate a relay coil.

Radio frequency and other techniques

The HEF 4046 PLL

This is a CMOS (complementary MOSFET transistor) logic device, which operates in a similar manner but needs more design effort to implement. Logic devices are actually part of Chapter 10, and it is there where the details of CMOS operation will be explained. For now, we can still make some useful headway by considering the functionality of the PLL.

The 4046 has the useful property that it can operate from a supply of 3–18 V, although the output capability is not as good as the 567. It is also electrostatic sensitive, so special precautions need to be taken when handling these ICs to prevent them from being damaged by small static shocks.

The 4046 is different in that it gives the choice of **two** phase comparators which perform similar, but subtly different, functions (Figure 9.4.5).

Figure 9.4.6 shows the difference between the outputs of the phase comparators.

Phase comparator 1 is a logic device called an exclusive-OR; as can be seen, the output is true whenever one OR other of the inputs

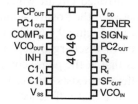

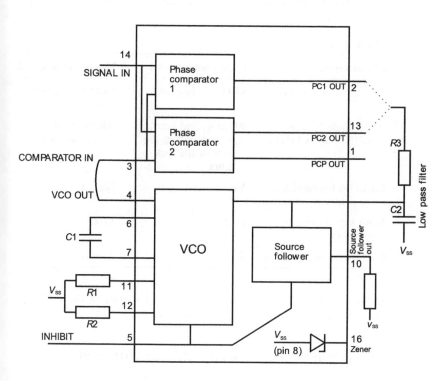

Figure 9.4.5 *The 4046 phase lock loop*

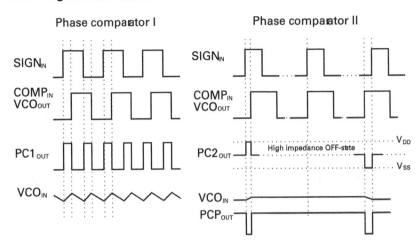

Figure 9.4.6 *The two 4046 phase comparators*

is true, but not both. The average voltage to the VCO input is supplied by the low-pass filter connected to the output of the phase comparator 1. This also causes the VCO to oscillate at the centre frequency (f_O).

Phase comparator 2 is an edge controlled circuit which is high impedance apart from the periods between the positive-going edges of the two input waveforms as shown in Figure 9.4.6. The low pass filter capacitor is thus either charged or discharged depending on which of the input waveforms arrives first at the phase comparator. This ensures that when a phase difference exists between the two inputs, the VCO will be driven to reduce that difference. When the PLL is in lock, the output PCP_{OUT} can be used to indicate that condition. Characteristics of comparator 1 and comparator 2 are shown in Table 9.4.1.

Table 9.4.1

Characteristic	Phase comparator 1	Phase comparator 2
No signal on $SIGN_{IN}$	VCO free runs at f_O	VCO free runs at f_{MIN}
Phase angle between $SIGN_{IN}$ and $COMP_{IN}$	90° at f_O and between 0° and 180° at the ends of the lock range	Always 0° in lock (+ve going edges)
Locks on harmonics of the centre frequency	Yes	No
Input signal noise rejection	High	Low
Capture frequency range	$f_C < f_L$	$f_C = f_L$

Designing a circuit with the 4046 requires the uses of the performance curves as shown in Figures 9.4.7 and 9.4.8.

The programmable frequency synthesizer

The phase locked loop acts in a way to make the frequencies at the two inputs of the phase comparator the same. Digital circuits exist

Radio frequency and other techniques 209

Figure 9.4.7 Values of R2/R1 as a function of f_{MAX}/f_{MIN}

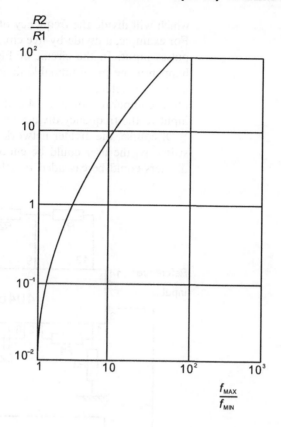

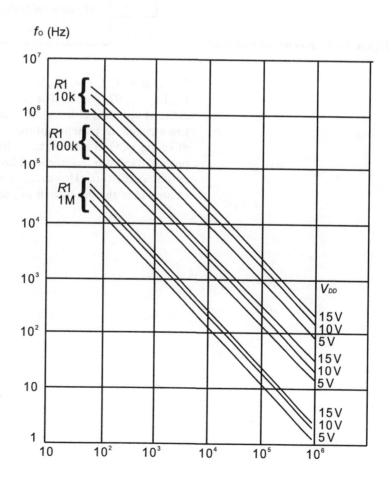

Figure 9.4.8 The relationship between R1, C1 and f_O

which will divide the frequency of its input by a specified amount. For example, a divide-by-five circuit will output 1 pulse for every 5 pulses input. In the circuit of Figure 9.4.9, consider what would happen if the programmable divider network was set up to divide by 5. If the reference input was 1 kHz, then for the two inputs to the phase comparator on pins 14 and 3 to have the same frequency, the input to the frequency divider would have to be 5 kHz.

Of course, the divider network could be controlled by a manual switch so the user could be choosing any number. Several of the dividers could be cascaded together so that instead of multiplying

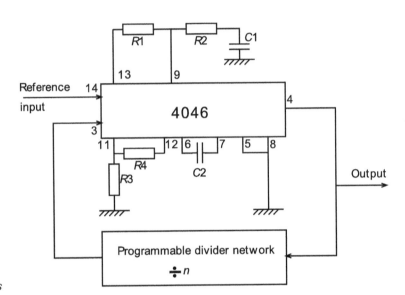

Figure 9.4.9 *Frequency synthesis*

the input by 1, 2, 3, 4, ..., 9, it could be multiplying the input by 11, 12, 13, ..., 97, 98, 99. The principle can be extended to as many digits as you like, although the circuit will get more complicated as you struggle to match capture and lock ranges and as you try to drive the VCO to even higher values. It would be vital to use a very precise oscillator reference. This might involve oven controlled crystal oscillators. However, it can be – and is – done as one technique in the generation of precise frequencies.

10 Logic circuits

Summary

Electronic logic is also known as **binary logic**. This reflects the fact that only two values are used. These are called '1' and '0' and are electronically represented by two voltage levels. The original transistorized logic circuit consisted of a simple amplifier which was allowed to go into saturation. The value h_{FE} describes the current gain of a common emitter transistor, but if more current is forced into the base so that the collector circuit cannot supply the supposedly required $h_{FE} \times I_B$ then the transistor is said to be saturated. In this state the collector–emitter voltage falls to somewhere in the range of 50–120 mV. The transistor is said to be fully ON. In this mode, it is used to switch on other systems. This is just the basis of a large family of logic devices which discriminate between two voltages LOW and HIGH. The gap between a voltage being allocated the tag '0' and the tag '1' is called the noise margin. The majority of logic devices use the nominal voltages 0 V and +5 V although modern systems operate from 0 V and 3.3 V. This reduces the power consumption and increases the speed of operation.

10.1 Logic gates and manipulation

Logic circuits are made of combinations of **logic gates** to perform a particular function. The main relationships are:

AND The output is true if all of the inputs are true.
OR   The output is true if one or more of the inputs are true.
NOT The output is true when the input is not true.

In Figure 10.1.1:

Lamp X will light if A AND B AND C are true.
Lamp Y will light if A OR B OR C is true.
Lamp Z will light if A is NOT true.

So, what is TRUE? A conventional logic family uses TRUE as being logic level 1 or ON or $+V$ and FALSE (NOT TRUE) as logic level 0 or OFF or 0 V. This is called **positive logic**.

Figure 10.1.1 *Switch logic elements*

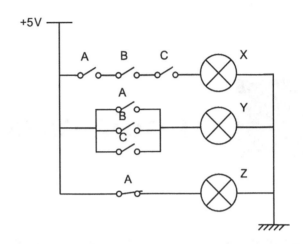

The opposite to these is called **negative logic** and is just as valid as positive logic, but uses the opposing conventions.

A logic circuit uses the basic AND, OR, and NOT relationships to combine one, two or more inputs to provide a logic output.

Logic symbols

There are two main families of logic symbols in common use:

MIL-STD-806   The so-called 'curvy' symbols.
IEC 617:12   The newer international electrotechnical 'rectangular' symbols.

These are summarized in Figure 10.1.2.

The MIL-STD-806 drawing standard identifies the logic function solely by its shape. Because they are easily interpreted by their shape alone, they are very popular with engineers and have not been entirely replaced by the IEC rectangular symbols.

The IEC 617:12 drawing standard uses rectangles which need a qualifying symbol inside. They form a powerful and comprehensive method of describing logical circuits and relationships. They are an effective method for producing schematic circuits which do not have a simple small scale integration (SSI) circuit. However, exactly *because* they are powerful, flexible and complex, they are not the easiest of circuits to use in other than their immediate simple logic gate form.

The Boolean expression uses a mathematical relationship named after George Boole, the British mathematician. It uses symbols to represent the logical functions of OR, AND, NOT and Exclusive-OR.

The **truth table** is another way of expressing logical relationships. It shows all the possible inputs and the corresponding outputs.

10.2 TTL

The NOT gate

TTL is **transistor–transistor logic** and reflects the internal structure of the device. The basic circuit is the NOT gate and is shown in Figure 10.2.1.

Transistor Q1 operates in what may be an unfamiliar mode. When the input is open circuit or the input is connected to the

Figure 10.1.2 Logic relationships, Boolean expressions and truth tables

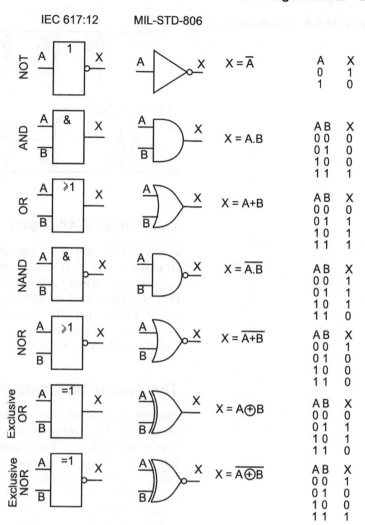

$+V_{CC}$ supply, then no current flows through the base–emitter junction. Instead, current flows through the base–collector junction to turn ON transistor Q2. This turns transistor Q4 ON and transistor Q3 OFF. The output is connected to the 0 V line via Q3 and hence the output is LOW at logic 0.

When the input is connected to the 0 V line, transistor Q1 turns ON conventionally, although there is no source of current through the collector of Q2. Hence, Q2 is OFF. This means that Q4 is OFF and Q3 is ON, making the output HIGH at a logic 1.

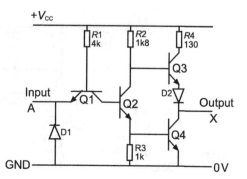

Figure 10.2.1 NOT gate

Figure 10.2.2 *The NAND gate*

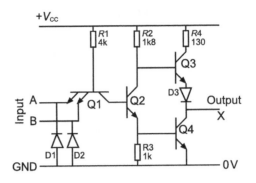

The 2-input NAND gate (Not-AND)

Figure 10.2.2 shows a 2-input NAND gate. It has the same structure as the NOT gate except that the input transistor Q1 has two emitters. When both of the inputs are HIGH at logic 1, Q1 is OFF and the output is LOW at logic 0. If either or both of inputs is pulled LOW, then the transistor Q1 starts to conduct and the output changes state to logic 1 or HIGH. An AND gate could be constructed from a NAND gate followed by a NOT gate.

The 2-input NOR gate (Not-OR)

The OR function of this circuit is achieved by the parallel action of Q2 and Q3. The operation is then explained in a similar manner to that of the NOT gate. The transistors Q1 and Q4 can only be turned ON by their respective inputs being pulled LOW by an input. If either A OR B is at a HIGH voltage level then Q2 or Q3 will be turned ON. Only when BOTH of the inputs are pulled LOW will neither of Q2/Q3 be turned on and the output will be 0 (LOW).

Note that for any TTL gate, the input must not be left unconnected. This is partly for pick-up/noise reasons, but mostly because an unconnected input is interpreted electrically as a $+V$.

Output circuits

The output stages in Figures 10.2.1, 10.2.2 and 10.2.3 are all the same. This is the so-called **totem-pole** output stage. One of the two output transistors is always ON to connect the output either to the $+V_{CC}$ line (1 output) or to the ground (0 output). One consequence of this is a **never: never** connect two totem-pole outputs together. If

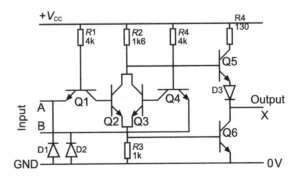

Figure 10.2.3 *The 2-input NOR gate*

Logic circuits

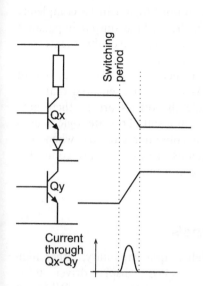

Figure 10.2.4 *Output transistor current flow*

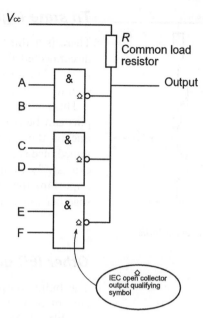

Figure 10.2.5 *Open collector outputs*

you do and they go to two different logic levels (one HIGH and one LOW) then there will be a low resistance path between V_{CC} and ground. One or both of the output stages will be destroyed.

The resistor in the totem-pole output is to reduce current flow between the $+V$ supply and the ground while the output transistors switch. A transistor cannot switch instantaneously between being ON and OFF. During the transition period, both transistors can be conducting for a very short period as shown in Figure 10.2.4. The resistor limits the current flow.

However, it is always good practice to install a $0.1\,\mu F$ or $0.01\,\mu F$ capacitor between V_{CC} and ground close to each IC. This minimizes effects on the power supply when the gates switch.

There is an alternative output – the open collector. In this version, only the lower output transistor of the totem-pole exists: Q4 for the NOT and NAND; Q6 for the NOR. The collector of the transistor is brought out as the output.

The outputs of open collector logic gates can be connected together as in Figure 10.2.5. If none of the output transistors is ON, then the output will be pulled high by the external resistor. If any one of the outputs is ON, then there is a current path to earth via the ON output stage, and so the output is LOW. For this reason, this type of circuit is called a WIRED-OR. The curly MIL-STD symbols do not have a special notation for open collector outputs. However, the IEC symbols have a qualifying symbol which is shown in Figure 10.2.5.

IEC qualifying symbols

Figure 10.2.6 shows a small selection of the special IEC symbols. The opposite of the open collector is shown; this is the open emitter. This is a less common implementation which is normally seen as a driver for display devices such as liquid crystals and light emitting diodes.

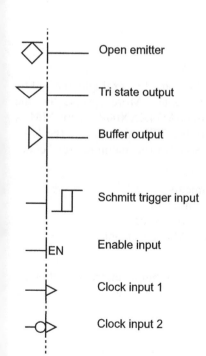

Figure 10.2.6 *IEC qualifying symbols*

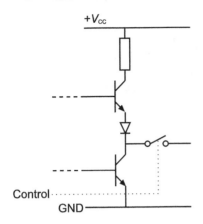

Figure 10.2.7 *Tri state logic schematic*

Tri state logic

There is a third logic state. If the output logic can be completely disconnected from the output pin, then a high output impedance results. The logic IC has a separate control pin to enable the tri state condition. It is equivalent to operating the switch in Figure 10.2.7.

This facility is one other situation where two or more logic output pins can be connected together. As long as it is ensured that only one of the outputs is active while the others are in their high impedance state, then no damage will result. Tri state logic is used extensively in microprocessor and computer circuitry when it is important that several devices such as memory ICs are connected together but that only one is active at a time.

Other IEC qualifying symbols

The buffer output indicates a high output capability, either high current or voltage. For example, a 7446 is a display driver and is capable of connecting its outputs to +30 V supplies. When an output goes LOW, it can sink up to 40 mA:

- Sinking occurs when an output takes current from an external device (in the same way that an open collector output would).
- Sourcing occurs when an output supplies current to an external device.

A Schmitt trigger input has hysteresis. This has been explained in Chapter 7, and means that the voltage level for switching into a logic 1 is different from that of logic 0.

The EN-able input is a separate control for other logic controls. The clock inputs will be explained shortly.

Families

Commercial TTL families have a 74 prefix. For example, the 7400 is an IC with four 2-input NAND gates. More specifically, the designation could be expanded to MM74XXX00P where MM is the manufacturer's code, XXX is the TTL structure of the logic circuit and P stands for Package. Some of the manufacturers' codes include:

- SN Texas Instruments
- CD Harris Semiconductor
- DM Fairchild Semiconductor
- M   SGS-Thomson Microelectronics
- MC Motorola

Initially, the XXX letters were not used, but as improvements were introduced the different variants of the functions had to be differentiated:

- L   Low power
- S   Schottky   higher speed
- LS   Low power Schottky   a combination of the above
- ALS Advanced low power Schottky

	Voltage range	Speed	Power
LS	+5 V 5%	10 ns	2 mW
ALS	+5 V 5%	7 ns	1 mW

(L and S are no longer in production.)

There are other subfamilies which belong to the CMOS group explained in Section 10.3.

Manufacturers use the location P to specify the type of package which holds this IC. It could be plastic or ceramic and be in one of several package sizes. The actual designation varies from manufacturer to manufacturer.

There is another prefix number, 54, which is a higher specification mostly used for military applications. The immediate differences between 74 and 54 devices are the operating temperature range and the supply voltage:

	Temperature range	Voltage supply tolerance
Commercial 74 families	0–70 °C	±5%
Military 54 families	−55–+125 °C	±10%

For most TTL families:

An input is recognized as a 1 if the input is >2 V
An input is recognized as a 0 if the input is <0.8 V

The noise immunity is the difference between these, i.e. 1.2 V:

A low output has a maximum of 0.2 V
A high output has a minimum of 3.3 V

Packages

There are hundreds of different variants of packages. Some of the simpler small scale integration (SSI) options include:

00	Quad 2-input NAND
02	Quad 2-input NOR
04	Hex inverter (NOT)
08	Quad 2-input AND
10	Triple 3-input NAND
11	Triple 3-input AND
20	Dual 4-input NAND
21	Dual 4-input AND
25	Dual 4-input NOR
27	Triple 3-input NOR
30	Single 8-input NAND
32	Quad 2-input OR
86	Quad 2-input Exclusive-OR
133	Single 13-input NAND

In addition to the stated number of logic gates in each IC, there will always be two other pins: V_{CC} and GND. In a 14 pin DIL, these will usually be $+V_{CC}$: pin 14 and GND: pin 7. In a 16 pin DIL these are usually $+V_{CC}$: pin 16 and GND: pin 8.

Summary of good practice when working with TTL

(1) Install a 0.1 µF or 0.01 µF capacitor between V_{CC} and ground close to each IC. This minimizes effects on the power supply when the gates switch.
(2) Tie unused inputs to ground or V_{CC} with a 1k resistor.
(3) Each TTL output can only drive up to 10 other TTL inputs. This is called its **fan-out**.

10.3 CMOS

CMOS stands for **complementary metal oxide semiconductor**. The word complementary refers to the fact that both p and n channel FETs are used. The original CMOS family carried the number sequence 4000. Unfortunately, the numbering is not the same as the 74xx family. For example:

4001	Quad 2-input NOR
4002	Dual 4-input NOR
4011	Quad 2-input NAND
4012	Dual 4-input NAND
4023	Triple 3-input NAND
4049	Hex inverter (NOT)
4068	Dual 4-input NAND
4070	Quad 2-input Exclusive-OR
4071	Quad 2-input OR
4072	Dual 4-input OR
4077	Quad 2-input Exclusive-NOR

The 4000 family members are slow, with a propagation delay of 50 ns, but the power consumption is very low at 1 µW. Another advantage is the wide operating voltage range. V_{CC} can be between 3 V and 18 V, which all makes them ideal for battery powered applications. The voltage threshold levels are different:

- **Input**
  Logic '1' is recognized above $\frac{2}{3} V_{CC}$
  Logic '0' is recognized below $\frac{1}{3} V_{CC}$
- **Output**
  The minimum logic '1' output is $V_{CC} - 0.01$ V
  The maximum logic '0' output is 0.01 V

This difference from TTL threshold levels, even when working from the same voltage supply, can cause problems when mixing the two technologies.

Figure 10.3.1 shows a comparison of the Quad 2-input CMOS and TTL pinouts.

The advantages of the standardization of the 74xx family were recognized, and the result was:

74HC - - High speed CMOS operating from 2 V to 6 V
74HCT - - High speed CMOS with TTL compatible supplies and inputs/output voltages

Figure 10.3.1 *Comparative pinouts*

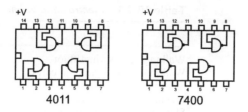

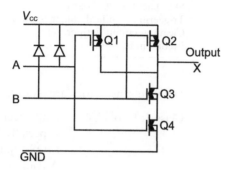

Figure 10.3.2 *A CMOS NAND gate*

74ACT - - Advanced CMOS with TTL standard voltage levels and pinouts

74AC - - Advanced CMOS with CMOS voltage level compatibility and TTL pinouts

In Figure 10.3.2, consider the FETs as perfect switches which are turned on and off by the gate voltage:

- A positive input voltage will turn an n-channel MOSFET ON (Q3 and Q4).
- A zero input voltage will turn a p-channel MOSFET ON (Q1 and Q2).

Q1 and Q2 are in parallel and so act as an OR gate, since if either one of them conducts then current will flow through to the output.

Q3 and Q4 are in series and so act as an AND gate since both of them need to conduct before the output is connected to the GND pin. This means that only if input A and input B are high will the output be 0. If either one of the inputs A and B are high ('1') then one or other of the n-channel devices will be off and one of the p-channel devices will be ON, connecting the output to V_{CC}. This gives rise to a NAND gate characteristic.

Electrostatic sensitivity

The p and n channel MOSFETs which are used in CMOS have very thin gate insulation layers – so much so that they can be destroyed by fairly low voltages. One of the precautions which must be taken with any MOS-based semiconductor is to take account of its vulnerability to static charges. Devices are routinely handled only in electrostatic device (ESD) protected areas. The semiconductors are transported in conductive tubes or conductive foam or conductive plastic bags which carry an ESD warning label. Table 10.3.1 shows everyday activities and the voltages which these can develop. They are more severe on dry days when the relative humidity is lower.

Table 10.3.1 *Electrostatic voltages and component sensitivities*

	70–90% r.h	10–20% r.h.
Walking across vinyl floor	250 V	12 kV
Walking across synthetic carpet	1.5 kV	35 kV
Sitting on a foam cushion	1.5 kV	18 kV
Picking up a standard plastic bag	600 V	20 kV
Sliding plastic box on a carpeted bench	1.5 kV	18 kV
Pulling tape from PC board	1.5 kV	12 kV
Skin packing PC board	3 kV	16 kV
Triggering standard solder remover	1 kV	8 kV
Cleaning circuit with eraser	1 kV	12 kV
Freon circuit spray	5 kV	15 kV

Damage sensitivities are as follows:

Class 1: 0 to 1 kV
- Unprotected MOS (discretes and ICs, especially VLSI)
- MOS capacitors (op amps; comparators)
- Advanced Schottky logic (FAST, ALS, LS)
- Junction FETs and low current (<0.15 A) SCRs
- Microwave and vhf transistors and ICs (especially Schottky)
- Precision (<0.5%) IC voltage regulators
- Precision (<0.1%) and low power (0.05 W) thin film resistors

Class 2: 1 to 4 kV
- MOS ICs with internal protection (CMOS, NMOS, PMOS)
- Schottky diodes (rectifiers)
- Linear ICs (bipolar)
- High speed bipolar logic (ECL, LS-TTL)
- Monolithic ceramic capacitors

Class 3: 4 to 15 kV
- Small signal diodes (<1 W) and transistors (<5 W)
- Low speed bipolar logic (TTL)
- Quartz and piezoelectric crystals

Guidelines for handling MOS devices:

- Store integrated circuits in conductive foam or conductive plastic storage tubes.
- Use a properly constructed static dissipating wrist strap and grounding lead.
- Use a properly grounded soldering iron and a grounded work bench.
- Do not apply signals to the input with the power off. (A low output impedance signal generator may forward bias the protection diodes and destroy the IC.)
- As with all integrated circuits, do not remove or insert MOS ICs with the power on.
- Connect all unused input leads to V_{CC} or V_{SS} (+V supply or ground). Never leave them open circuit because they will accumulate static and float to unpredictable (or damaging) voltage levels.
- Put resistors in series with any MOS inputs connected to points off a printed circuit board.

As with TTL, do not tie the outputs of CMOS circuits together unless they are open-drain or tri-state logic with outputs which are not simultaneously activated.

10.4 Combinational logic circuits

The preceding sections covered the technology and theory of digital circuits. This section covers some simple applications.

Inputs

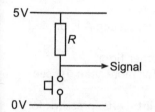

Figure 10.4.1 *A simple switch input*

Inputs can be from compatible sources or from any of a multitude of switch type inputs. When connecting a switch to a logic circuit, especially TTL, there are a few precautions to take. Switches are normally connected in a pull down configuration as shown in Figure 10.4.1. The resistor R (typically 4k7) normally references the TTL input to the $+V_{CC}$, so the input is interpreted as a '1'. Pressing the button pulls the voltage down to 0 V and so the input is interpreted as a '0'.

With TTL, the switch is rarely configured as a pull-up. If you try to do this, the resistor R would be connected from the TTL input to ground. Hopefully, this might be interpreted as a '0', but if the resistor is large enough, then a voltage would be developed across R. If the TTL source current is large enough, then it could even cross the '1' voltage threshold (2 V).

Example 10.4.1

A burglar alarm system has a master switch A to switch the system on, a switch B on the window and a switch C on the door. Devise a logic network (to MIL-STD-806) to ring the bell when the master switch is on and when either the window or the door switches are on.

A	B	C	Bell
0	0	0	0
0	0	1	0
0	1	0	0
0	1	1	0
1	0	0	0
1	0	1	1
1	1	0	1
1	1	1	1

The Boolean expression describing this situation would be:

$$\text{Bell} = A \cdot (B + C)$$

A suitable circuit would be as shown in Figure 10.4.2.

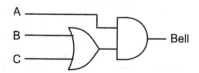

Figure 10.4.2 *The burglar alarm solution*

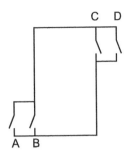

Figure 10.4.3

Problems 10.4.1

(1) Figure 10.4.3 shows a dust-free area with two entrances. Each entrance has two doors for which a warning bell must ring if both doors A and B or doors C and D are open at the same time. State the Boolean expression depicting this occurrence and devise a logic network to operate the bell (drawn to IEC standards).

(2) Lift doors should close if the master switch A is on and either:
 - a call B is received from any other floor; or
 - the doors C have been open for more than 10 seconds; or
 - the selector push with the lift D is pressed for another floor.

Devise a logic circuit to meet these requirements (MIL-STD-806). Assume that the timing constraints of the doors C are handled separately and you have a logic level which goes from 0 to 1 whenever the doors have been open for more than 10 s.

(3) A logic signal is required to give an indication when:
 - the supply to an oven is on; and
 - the temperature of the oven exceeds 210 °C; or
 - the temperature of the oven is less than 190 °C.

Show the Boolean expression required to describe the solution and draw the IEC617:12 solution to the problem.

(4) A water tank feeds three separate processes. When any two of the processes are in operation at the same time, a signal is required to start a pump to maintain the head of water in the tank. Devise a logic circuit to solve the problem and state its Boolean expression.

(5) The circuit of Figure 10.4.4 shows the CMOS version of a NOR gate. Explain how it works.

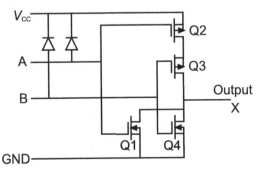

Figure 10.4.4 *The CMOS NOR gate*

10.5 Minimization

So far, we have only looked at very simple analyses of a given situation and tried to apply some thought to the process of determining the Boolean equation and/or the logic diagram from the truth table. The truth table is really the starting point of the solution. Once this is obtained, then a solution can always be worked out. However, it may not always be the most efficient or minimal solution. For this reason, various minimization techniques were evolved. These will be covered briefly, but most practical problems (i.e. non-trivial) are solved using one or other of the many software packages. Before these were available, engineers had to get down to minimization with Boolean logic or visual mapping techniques such as those developed by Karnaugh, or algebraic methods such as those developed by Quine and McCluskey. Such solutions are effective, but labour intensive.

Boolean Logic minimization uses combinations of reduction formulae from the available logic relationships:

$$A.A = A \qquad A.0 = A$$
$$A./A = 0 \qquad A + 0 = A$$
$$A + A = A \qquad A.B = B.A$$
$$A + /A = 1 \qquad A + B = B + A$$
$$A.1 = A \qquad A.(B+C) = A.B + A.C$$
$$A + 1 = 1 \qquad A + B.C = (A+B).(A+C)$$

Boolean logic minimization is assisted by another set of rules which were set out by De Morgan. These allow the conversion of one form of Boolean expression into another which may be easier to manipulate:

$$\overline{A} + \overline{B} = \overline{(A.B)} \qquad \overline{(A+B)} = \overline{A}.\overline{B}$$

An expression of De Morgan's theorem in words would be 'replace all ANDs with ORs; all ORs with ANDs; negate each term; negate each expression'.

For example:

$$A.\overline{B} + C \text{ would convert to } \overline{(\overline{A} + B)}.\overline{C}$$

$$\overline{(A + \overline{B})} \text{ would convert to } \overline{A}.B$$

There are two ways in common use of stating a logic equation:

$$\text{Product of sums} \quad (A+B).(C+D)$$
$$\text{Sum of products} \quad E.F + G.H$$

This confusingly relates the OR function with its algebraic symbol '+' and the AND function with its algebraic symbol '.'. De Morgan's theorem allows conversion from one to the other.

The Karnaugh map is like a two-dimensional truth table with the input variables arranged according to a Gray code – i.e. only one bit changes at a time (Figure 10.5.1).

The '1' variables are enclosed by loops. The loops must enclose 2, 4 or 8 adjacent 1s. In Figure 10.5.1, the output states would be

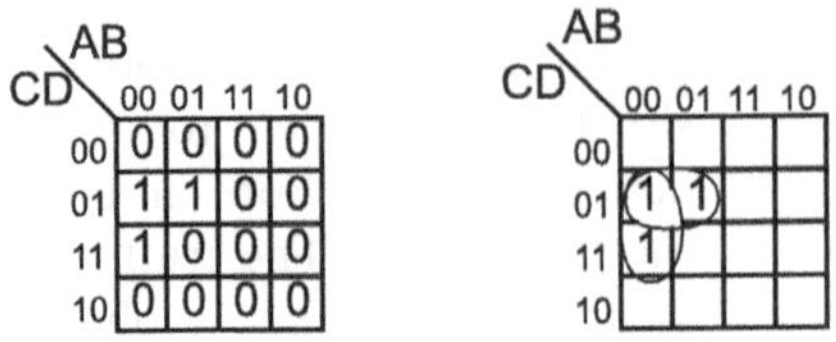

Figure 10.5.1 *Karnaugh maps*

$\overline{A}.\overline{B}.D$ for one loop and $\overline{A}.\overline{C}.D$ for the other. (The AB axis showing a 1 in the 00 column implies the condition that there is a 1 when there is NOT-A and NOT-B). The vertical loop has C as a 'don't care' term since the loop covers C values of 0 and 1 – i.e. C can take any value without affecting the output of that term. The horizontal loop has B as a don't care term since the loop covers B values of 0 and 1.

The full output is $X = \overline{A}.\overline{B}.D + \overline{A}.\overline{C}.D$. There will be as many output terms as there are loops. The minimized output expression is achieved when every 1 in the map has been looped at least once.

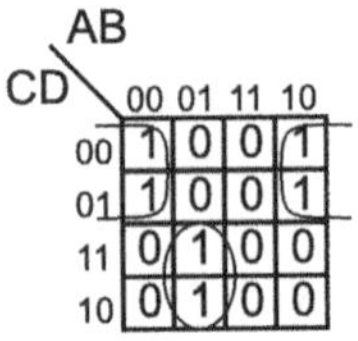

Loops wrap around at the edge as shown in Figure 10.5.2. If there is a 1 in each corner position, then this is considered to be a single loop encompassing four terms. The output in this case is minimized to $X = \overline{B}.\overline{C} + \overline{A}.B.C$, corresponding to the coverage of each loop.

Figure 10.5.2

The Quine–McCluskey technique and its derivatives are a manual, non-table driven version of the Karnaugh map method. It involves pattern matching between adjacent lines and as such is very suited to computerization and software solution. Before minimization, the expression must be put into its canonical form whereby it is expanded so that there is a sum of products or product of sums, each term of which contains only one incidence of the variable. For example, the expression:

$$X = (A + \overline{A}).\overline{B}.\overline{C} + (B + \overline{B}).A.\overline{C}$$

could be converted using the rules above to the canonical form of

$$X = A.B.\overline{C} + A.\overline{B}.\overline{C} + \overline{A}.\overline{B}.\overline{C}$$

The rest of the method relies on creating tables of results which compare all the expression in the canonical form.

Problems 10.5.1

(1) Using the Boolean equations and De Morgan's theorem, simplify:
  (a) $X = A + B + B.C$
  (b) $X = \overline{A.B + B + B.\overline{C}}$
  (c) $X = (A + B).(\overline{A} + C).(A + B)$

(2) Derive the truth tables of:
  (a) $X = A.B + \overline{B}.C + A.C$
  (b) $X = A.(\overline{B} + \overline{C})$
  (c) $X = B.C + \overline{A}.B$

(3) Minimize the Karnaugh map of Figure 10.5.3 and draw the resulting logic circuit diagram. A more minimal solution will result if this is considered as two blocks

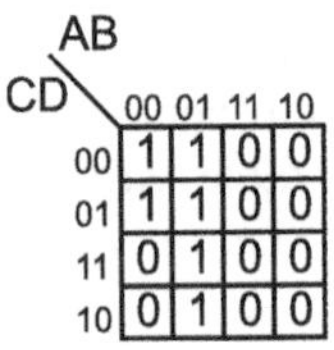

Figure 10.5.3

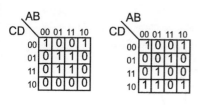

Figure 10.5.4   **Figure 10.5.5**

of 4 rather than a block of 4 and a block of 2. Try both ways and see.

(4) Minimize the Karnaugh map of Figure 10.5.4 and draw the resulting logic circuit diagram.
(5) Minimize the Karnaugh map of Figure 10.5.5 and draw the resulting logic circuit diagram.

10.6 Programmable logic devices

A programmable logic device is a general name for a logic device for which the user can specify the logical relationship between the inputs and outputs. Fundamentally, these devices consist of arrays of AND gates, OR gates and inverters which can be connected together in many possible ways (simple combinatorial logic). Devices can also incorporate latches (registered logic). The arrangement is generally that the inputs feed into an array of AND gates, and the output of all the AND gates is fed into a set of OR gates. To start with, all the inputs feed into all the AND gates, but these devices are programmable, so it is possible selectively to disconnect some of the inputs if necessary. There are several methods of storing the information about which connections are left intact and which disconnected.

Programming was originally achieved by blowing miniature nickel-chromium (Ni-Cr) fuses integrated onto the silicon 'chip'. The method is still used for some applications, but more recent devices mostly rely on MOS transistors with an embedded gate which is isolated within the silicon substrate. The gate of these transistors can have a charge stored on it by a programming procedure which applies a local high voltage near the gate and electrons which are raised to a higher energy level can 'tunnel' across to the gate. This charge will stay stored since there is no conduction path away from this gate (although some programmable logic devices have a mechanism whereby the charge on the gate can be discharged by exposure to ultraviolet light or by an electrical erasure procedure). The storage transistor helps determine whether or not connections are made within the logic array.

The symbols used in some of these devices are illustrated in Figure 10.6.1. Figure 10.6.1(a) shows the programmable-AND array logic equivalent. Note that each input passes through a buffer-inverter which produces not only a copy of the input, but also its complement. Thus, it is possible to not-connect, or to connect the true or the complement of any input to the AND gate.

Figure 10.6.1(b) shows how this circuit is represented in a manufacturer's typical notation. The AND-gate inputs are represented by a single line, commonly described as the 'product term'. All array inputs (true and complement) are shown connected to a single input AND gate. For example, the AmPAL16L8 has 16 inputs, but all of these inputs and their complements are routed to each AND gate. The fuse states, either left intact or blown, determine the customized function of the device.

Figure 10.6.2(a) shows how the AND array is connected through the fuse arrangement into the array of OR gates while Figure 10.6.2(b) shows the conventional notation which is used by most manufacturers.

Figure 10.6.1 *The AND array*

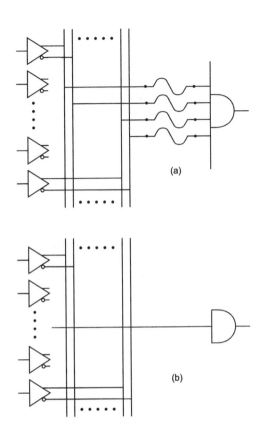

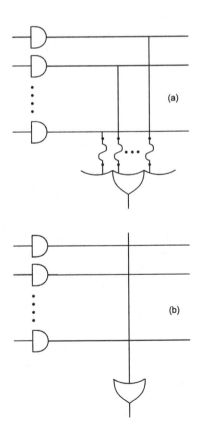

Figure 10.6.2 *The OR array*

Figure 10.6.3 *A simple function*

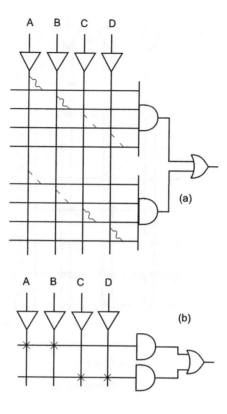

Figure 10.6.3(a) shows the fuse implementation of the function (A.B + C.D). In the diagram, you can see which fuses have had to be blown in order to complete the function. Figure 10.6.3(b) shows how the function might be represented using the more compact notation.

There are several types of PLD notation, summarized as:

PLD	Programmable logic device
PAL	Programmable array logic
PLA	Programmable logic array
PROM	Programmable read only memory

The arrangement is consistent in that they are constructed so that the inputs meet a first level of an AND array and then a second level of an OR array. The differences between these are:

	PAL	PLA	PROM
AND	Programmable	Programmable	Fixed
OR	Fixed	Programmable	Programmable

A PAL consists of a programmable AND array with true and inverting inputs to each AND gate, coupled to one fixed OR array per output (Figure 10.6.4). PAL is the most useful when designing AND intensive circuits.

A PLA (sometimes called a field-programmable logic array) is a truly flexible PLD, which can be complex to work with (Figure 10.6.5). They have not been as popular as PALs, possibly because of the longer propagation delays which follow from the extra control circuitry needed to implement the additional flexibility.

Figure 10.6.4 *The PAL*

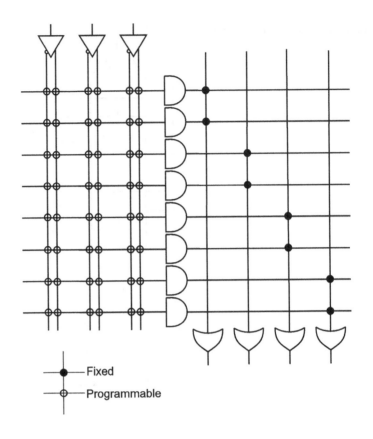

Figure 10.6.5 *The PLA*

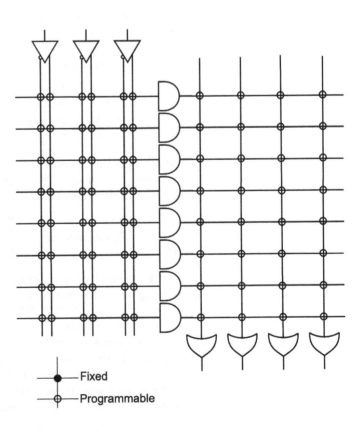

Figure 10.6.6 *The PROM*

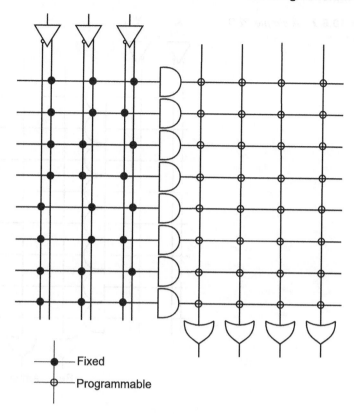

A PROM is a PLD which is used as a memory device for microprocessor based systems, although it does find an important use in simple PLD applications (Figure 10.6.6). It is characterized by a fixed AND array coupled to a programmable OR array. Every possible combination of binary inputs is provided and the OR array can be programmed to suit. The 'E' in front of EPROM or EPLD implies that the device is erasable by one or other of the technologies.

Problems 10.6.1

(1) It is necessary to design a traffic light controller which has as its inputs a binary clock in the sequence 00 ... 01 ... 10 ... 11 ... 00 ... etc. The outputs are to follow the normal UK traffic light sequence of Red, Red + Amber, Green, Amber, Red, etc. Devise a set of Boolean equations for each traffic light colour and identify which of the fuses on the diagram below should be blown.

INPUT		OUTPUT		
X	Y	Red	Amber	Green
0	0			
0	1			
1	0			
1	1			

Figure 10.6.7 *A simple PLD structure*

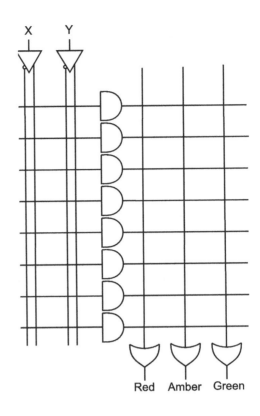

Equations:

Red =
Amber =
Green =

(2) The aim is to provide a PLD which will provide the necessary logic functions for an electronic die. The input is a binary clock which will provide the count sequence 000 ... 001 ... 010 ... 011 ... 100 ... 101 ... 000 ... etc. From this, the appropriate LEDs should be driven:

```
    f           a
  e     g     b
    d           c
```

Complete the following table, substituting 1s for the occasion on which the LEDs will be lit:

	x	y	z	a	b	c	d	e	f	g
1	0	0	0							
2	0	0	1							
3	0	1	0							
4	0	1	1							
5	1	0	0							
6	1	0	1							

Produce the Boolean equations for each of the LEDs:

Figure 10.6.8

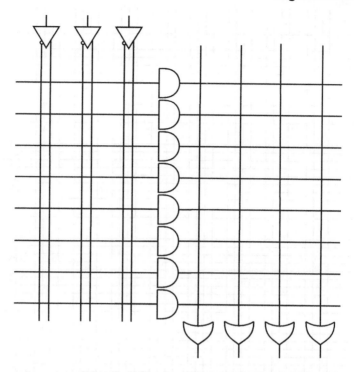

a =
b =
c =
d =
e =
f =
g =

Now design the PLD and show which fuses ought to be left intact on the diagram of Figure 10.6.8.

Figures 10.6.9 and 10.6.10 show two diagrams of practical PAL devices. The first is the 16L8 and the second is the 16R8. Both have 16 inputs and 8 outputs and represent the very low end of the capability of current devices. They are included because they are among the easiest to understand. With the 16 inputs and 8 outputs, it does not mean that there are a total of 24 data pins, since a close examination will show that some pins could be used as either an input or an output. The first diagram is a combinational device only while the other is a registered design with output latches.

The advantages of PLDs are:

- **Ease of design/implementation**
  - The designer can design, simulate and create hardware solutions.
  - CMOS erasable devices can be reprogrammed if an error is discovered.
  - Board layout is simplified (the designer can change which pins perform specific functions in order to cut down on PCB track crossovers).

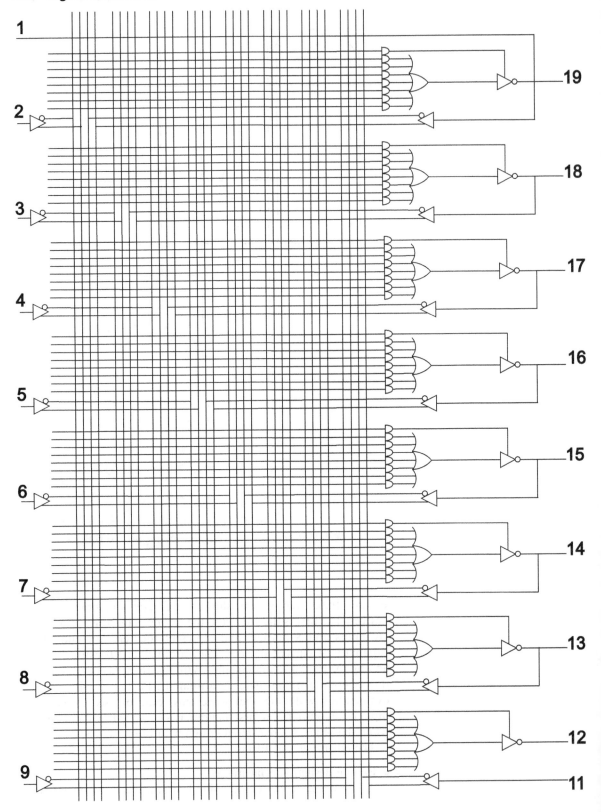

Figure 10.6.9 *The 16L8 PAL*

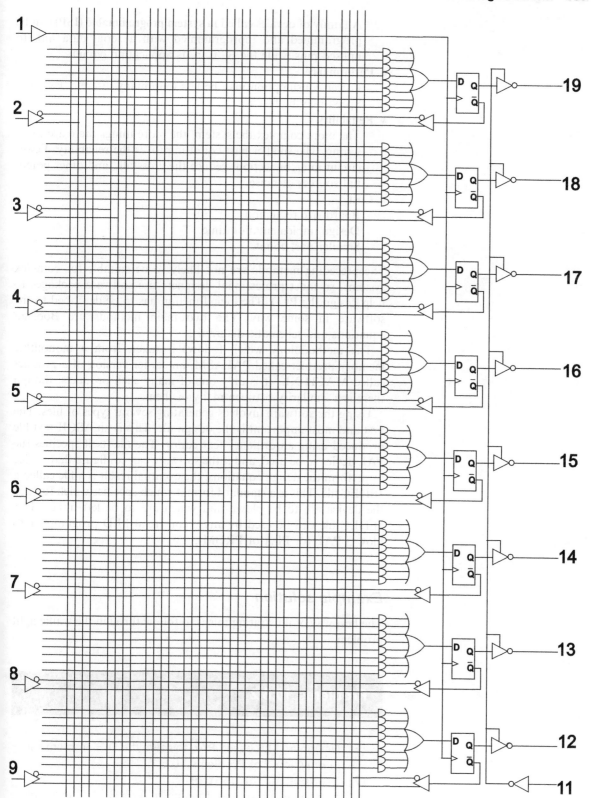

Figure 10.6.10 *The 16R8 PAL*

- A family of devices called **in system programmable** (ISP) can be programmed and reprogrammed with a serial data link to unused pins while the IC is in its application circuit.
- **Performance**
  - Power consumption may be less than many discrete devices.
  - Speed may be faster.
- **Reliability**
  - The more complex the system, the more things there are to go wrong. The smaller part count improves the reliability. Fewer circuit interconnections reduce the crosstalk, and other potential sources of noise.
- **Cost**
  - Board size reduction.
  - Design/implementation time.
  - Smaller part count.

The preceding methods are not likely to be suitable for complex systems, so most professional designers use software packages to help design the PLD. The logic function has to be described to the software in some way. The two main methods are Boolean equations and schematic diagrams.

The methodology which follows is taken from a monolithic memory product called PALASM, an earlier version of a product called MACH XL. It is only one of many such products, though there is a similarity between many of them.

Using the software involves generating several types of files. The first and most important is the design file. This is an ASCII text file produced using an ordinary text editor. The file itself specifies the PAL device to be used, and if the software can configure that device to perform the design equations, the eventual result is a file called a JEDEC file. This is a standard format which specifies which fuses in the device are to be blown and which are to be left intact. This JEDEC file can be used by any commercial PLD programmer to blow a PAL (or PLA or PROM).

Example solution

The best way of illustrating this is to follow through the traffic light problem.

Problems 10.6.2

(1) Design a 16L8 based PAL to solve the DIE problem set in question 2 of Problems 10.6.1.

(2) **Stepper motor control**
A stepper motor has four coils A B C D. The input clock cycles through the sequence 000 001 010 011 100 101 110 111 etc. In response, the coils must be energized in the sequence A A+B B B+C C C+D D D+A. Derive a set of Boolean equations and hence design the application for a 16L8 PAL.

The headings seen here are mandatory, i.e. they must head up each and every design file. The key words are compulsory (the program looks for them) but the actual text supplied alongside is for the designer's own reference purposes.

This section uses the keyword CHIP to declare which device is actually to be used. The word between CHIP and the device is for the designer's own use. Following the CHIP declaration is the PIN list which gives a user label for each of the pins. All pins must be accounted for and the pins must be presented in the correct numerical order. Two reserved names are GND and VCC. When allocating names, you must refer to the CHIP layout to check on the function of the various pins. The only pins used in this exercise are 2, 3, 10, 12, 13, 14, 20.

The EQUATIONS section comes next. It defines the Boolean relationship between the inputs and the outputs. The 'bar' symbol is the Boolean negation symbol.

The SIMULATION section is optional. It allows you to check the validity (or otherwise) of your design. As can be seen, the trace is following the status of C0, C1, RED, AMB, GRN as the state of C0 and C1 are varied.

```
TITLE        Traffic Light Controller
PATTERN      MJ001
REVISION     A
AUTHOR       Mike James
COMPANY
DATE         30/2/99

CHIP traffic    PAL16L8

; PINS 1 2 3 4 5 A C1 C0 B D
; PINS 6 7 8 9 10 E F G H GND
; PINS 11 12 13 14 15 I RED AMB GRN J
; PINS 16 17 18 19 20 K L M N VCC

EQUATIONS

RED = $\overline{C1}$
AMB = C0
GRN = C1 * $\overline{C0}$

SIMULATION
TRACE_ON C1 C0 RED AMB GRN
SETF $\overline{C0}$ $\overline{C1}$
SETF C0 $\overline{C1}$
SETF $\overline{C0}$ C1
SETF C0 C1
SETF $\overline{C0}$ $\overline{C1}$
TRACE_OFF
```

(3) **New logic device (1)**
This to be a 3 input, 4 output binary logic device. The outputs are to represent the number of 1s present at the input. For example, an input of 101 would make the '2' output active. Devise the truth table and logic equations for the outputs. Implement the design for a 16L8 PAL.

(4) **New logic device (2)**
This is to be a 3 input, 2 output status indicating circuit. The outputs indicate (in binary) the number of 1s in the input. For example, the input 101 would result in an output of 10. Devise a truth table and logic equations for this circuit. Implement the design for a 16L8 PAL.

(5) **Odd and even indicator**
The input is to be one NIBBLE of data while the output is to be the seven-segment code for O or E. This is a device which indicates whether an odd or an even number of 1s are present in the input. Create a truth table and a set of Boolean equations for each segment.

(6) **Interrupt priority encoder**
This system is to provide an indication of the priority of interrupts for a microprocessor-based system. There are to be 4 active high inputs called INT0 to INT3 and

The file was given the name of TRAFFIC. The source file (the program just listed) was saved in a file called traffic.pds. As a result of running the various procedures, the following files were generated:

traffic.pds	the source file
traffic.xpt	a fuse plot file (i.e. which fuses are to be blown); this is shown opposite
traffic.jed	the formal JEDEC file required by PAL programmers
traffic.hst	a full simulation history
traffic.trf	a simulation history based on the traced inputs/outputs
traffic.jdc	a composite JEDEC and history file

The simulation file traffic.trf was:

PALASM SIMULATION, V2.23D – MARKET RELEASE (2-14-89) (C) – COPYRIGHT ADVANCED MICRO DEVICES INC., 1989
PALASM SIMULATION SELECTIVE TRACE LISTING

Title:	Traffic Light Controller
Author:	Mike James
Pattern:	MJ001
Company:	
Revision:	A
Date:	30/2/99

PAL16L8
TRAFFIC
Page: 1

```
ggggg
C1      LLHHL
C0      LHLHL
RED     HHLLH
AMB     LHLHL
GRN     LLHLL
```

PALASM PAL ASSEMBLER V2.23D – MARKET RELEASE (2-14-89) (C) – COPYRIGHT ADVANCED MICRO DEVICES INC., 1989

TITLE:	Traffic Light Controller
AUTHOR:	Mike James
PATTERN:	MJ001
COMPANY:	Yeovil College
REVISION:	A
DATE:	30/2/99

PAL16L8
TRAFFIC

```
         11   1111 1111 2222 2222 2233
    0123 4567 8901 2345 6789 0123 4567 8901
 0  XXXX XXXX XXXX XXXX XXXX XXXX XXXX XXXX
 1  XXXX XXXX XXXX XXXX XXXX XXXX XXXX XXXX
 2  XXXX XXXX XXXX XXXX XXXX XXXX XXXX XXXX
 3  XXXX XXXX XXXX XXXX XXXX XXXX XXXX XXXX
 4  XXXX XXXX XXXX XXXX XXXX XXXX XXXX XXXX
 5  XXXX XXXX XXXX XXXX XXXX XXXX XXXX XXXX
 6  XXXX XXXX XXXX XXXX XXXX XXXX XXXX XXXX
 7  XXXX XXXX XXXX XXXX XXXX XXXX XXXX XXXX
 8  XXXX XXXX XXXX XXXX XXXX XXXX XXXX XXXX
 9  XXXX XXXX XXXX XXXX XXXX XXXX XXXX XXXX
10  XXXX XXXX XXXX XXXX XXXX XXXX XXXX XXXX
11  XXXX XXXX XXXX XXXX XXXX XXXX XXXX XXXX
12  XXXX XXXX XXXX XXXX XXXX XXXX XXXX XXXX
13  XXXX XXXX XXXX XXXX XXXX XXXX XXXX XXXX
14  XXXX XXXX XXXX XXXX XXXX XXXX XXXX XXXX
15  XXXX XXXX XXXX XXXX XXXX XXXX XXXX XXXX
16  XXXX XXXX XXXX XXXX XXXX XXXX XXXX XXXX
17  XXXX XXXX XXXX XXXX XXXX XXXX XXXX XXXX
18  XXXX XXXX XXXX XXXX XXXX XXXX XXXX XXXX
19  XXXX XXXX XXXX XXXX XXXX XXXX XXXX XXXX
20  XXXX XXXX XXXX XXXX XXXX XXXX XXXX XXXX
21  XXXX XXXX XXXX XXXX XXXX XXXX XXXX XXXX
22  XXXX XXXX XXXX XXXX XXXX XXXX XXXX XXXX
23  XXXX XXXX XXXX XXXX XXXX XXXX XXXX XXXX
24  XXXX XXXX XXXX XXXX XXXX XXXX XXXX XXXX
25  XXXX XXXX XXXX XXXX XXXX XXXX XXXX XXXX
26  XXXX XXXX XXXX XXXX XXXX XXXX XXXX XXXX
27  XXXX XXXX XXXX XXXX XXXX XXXX XXXX XXXX
28  XXXX XXXX XXXX XXXX XXXX XXXX XXXX XXXX
29  XXXX XXXX XXXX XXXX XXXX XXXX XXXX XXXX
30  XXXX XXXX XXXX XXXX XXXX XXXX XXXX XXXX
31  XXXX XXXX XXXX XXXX XXXX XXXX XXXX XXXX
32  XXXX XXXX XXXX XXXX XXXX XXXX XXXX XXXX
33  XXXX XXXX XXXX XXXX XXXX XXXX XXXX XXXX
34  XXXX XXXX XXXX XXXX XXXX XXXX XXXX XXXX
35  XXXX XXXX XXXX XXXX XXXX XXXX XXXX XXXX
36  XXXX XXXX XXXX XXXX XXXX XXXX XXXX XXXX
37  XXXX XXXX XXXX XXXX XXXX XXXX XXXX XXXX
38  XXXX XXXX XXXX XXXX XXXX XXXX XXXX XXXX
39  XXXX XXXX XXXX XXXX XXXX XXXX XXXX XXXX
40  ---- ---- ---- ---- ---- ---- ---- ----
41  -X-- ---- ---- ---- ---- ---- ---- ----
42  ---- X--- ---- ---- ---- ---- ---- ----
43  XXXX XXXX XXXX XXXX XXXX XXXX XXXX XXXX
44  XXXX XXXX XXXX XXXX XXXX XXXX XXXX XXXX
45  XXXX XXXX XXXX XXXX XXXX XXXX XXXX XXXX
46  XXXX XXXX XXXX XXXX XXXX XXXX XXXX XXXX
47  XXXX XXXX XXXX XXXX XXXX XXXX XXXX XXXX
48  ---- ---- ---- ---- ---- ---- ---- ----
49  ---- -X-- ---- ---- ---- ---- ---- ----
50  XXXX XXXX XXXX XXXX XXXX XXXX XXXX XXXX
51  XXXX XXXX XXXX XXXX XXXX XXXX XXXX XXXX
52  XXXX XXXX XXXX XXXX XXXX XXXX XXXX XXXX
53  XXXX XXXX XXXX XXXX XXXX XXXX XXXX XXXX
54  XXXX XXXX XXXX XXXX XXXX XXXX XXXX XXXX
55  XXXX XXXX XXXX XXXX XXXX XXXX XXXX XXXX
56  ---- ---- ---- ---- ---- ---- ---- ----
57  X--- ---- ---- ---- ---- ---- ---- ----
58  XXXX XXXX XXXX XXXX XXXX XXXX XXXX XXXX
59  XXXX XXXX XXXX XXXX XXXX XXXX XXXX XXXX
60  XXXX XXXX XXXX XXXX XXXX XXXX XXXX XXXX
61  XXXX XXXX XXXX XXXX XXXX XXXX XXXX XXXX
62  XXXX XXXX XXXX XXXX XXXX XXXX XXXX XXXX
63  XXXX XXXX XXXX XXXX XXXX XXXX XXXX XXXX
```

SUMMARY

TOTAL FUSES BLOWN = 220

there should be 3 outputs. One of these is an active low interrupt request line and the other two should indicate the priority of the interrupt. If any input goes high, the INT output should go low. The binary outputs should then indicate the priority of the input — i.e. if INT1 becomes active the output should be 01. If INT1 and INT2 become active, then the binary output should be 10.

(7) **Switch indicator**

The input is to be 4 logic level switches, while the output should be a 'switch pressed' indicator and two binary outputs which indicate the identity of the pressed switch. If two or more input switches are pressed, then the 'switch pressed' indicator should not be active.

(8) **4-bit gray to binary converter**

This design for a 16L8 PAL should convert the Gray rotary code to true binary. A Gray code is one where only one bit changes in going from one state to the next. A four-bit sequence is:

```
0000
0001
0011
0010
0110
0111
0101
0100
1100
1101
1111
1110
1010
1011
1001
1000
```

11 Sequential logic

Summary

Logically enough, sequential logic is where a circuit proceeds through a set of logic states according to the rate of a controlling clock and according to the inputs and/or outputs of the system.

There are two methods of developing sequential solutions to given problems. These are loosely called the informal and formal methods. Informal is an intuitive technique which relies on modelling or construction to verify the correct operation. Formal methods produce an output which can reliably be predetermined by the development method in use.

11.1 Simple latches

In the circuit of Figure 11.1.1, the output of each NAND gate is fed back to the input of the other. This cross-coupled gate is the basis of a latch. There are two complementary outputs which are called Q and $\overline{Q}$. The inputs are normally at a logic 1. The two inputs are called $\overline{SET}$ and $\overline{RESET}$. When the $\overline{SET}$ input is changed to a 0, gate A output becomes 1 which when fed back to the input of gate B changes the output of B to 0. This is called 'toggling' the latch. One of the inputs of gate A is now at 0, so further 1–0–1 excursions of the $\overline{SET}$ input will have no effect. To reset the latch, the $\overline{RESET}$ line must be cycled through the sequence 1–0–1. The circuit would also work in a similar way if cross-coupled NOR gates were to be used instead.

Figure 11.1.2 shows a simple modification which normalizes the inputs to be a SET and RESET input. That is, a positive logic control.

Figure 11.1.3 shows a SET–RESET latch that has two novel features:

(1) An asynchronous clock which controls the output according to the state of the SET or RESET input. The output will only change when the input is applied:

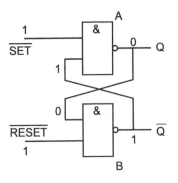

Figure 11.1.1 *The simplest SET–RESET latch*

Sequential logic 239

Figure 11.1.2 *Modified input to the SET–RESET latch*

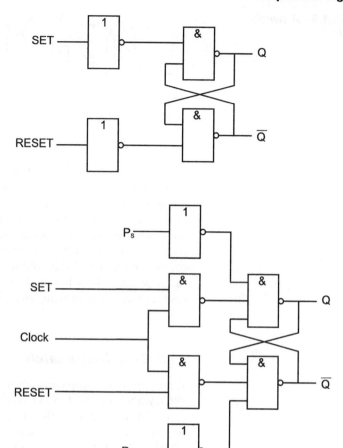

Figure 11.1.3 *Clocked SET–RESET latch*

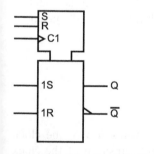

Figure 11.1.4 *An IEC SR flip-flop*

S	R	Output after the clock pulse
0	0	Unchanged
0	1	0
1	0	1
1	1	Undefined

(2) An asynchronous SET and RESET control. The SET control is sometimes referred to as a PRESET input and the RESET as a CLEAR. The IEC 617:12 symbol for this could be described in Figure 11.1.4.

The rectangle is the usual symbol with the letters S and R defining the SET–RESET operation. The T-shape on the top of the rectangle is the control block. There would be three controls in this instance. The clock input with the identifying number 1 qualifies the S and R inputs in the main box. In the logic symbols, a number after a control of any type refers to or is qualified by the input or output with the same number. The asynchronous control would affect the output directly, irrespective of the state of the clock. In this symbol, the alternative form of the negative logic designation is used. This is the angled line instead of the negation circle. The S–R latch is not normally found in many circuits. The most common use of the S–R latch is as a switch debouncer for logic circuits (Figure 11.1.5).

Figure 11.1.5 *A switch debouncer*

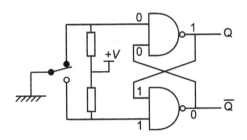

When a mechanical switch operates, the moving contact is likely to bounce against the fixed contact for a period of up to 50 ms. In doing so, it could generate many switching signals, and if it was part of a counting circuit, many false counts would be generated. When the bounce occurs, the moving part only bounces against the new contact. It does not bounce between the two contacts. When the contact moves over, the output toggles at the first bounce against the new edge. It will not change back again until the switch is operated and the moving element first bounces against the new contact.

11.2 The D-type latch

A more practical implementation is the D-type latch. It is given the IEC symbol and MIL-STD symbol as shown in Figure 11.1.6.

Specifically, this is the TTL dual D-type latch 74LS74. In practical terms, when the clock input changes from a 0 to a 1, whatever logic state on the D input is transferred to the Q output. The $\overline{Q}$ merely reflects the complement of the Q output. This presupposes that the Preset and Clear are both at 1. If either move to the 0 condition, then the stated function is enacted, regardless of the state of the clock. Its truth table is usually given as:

D	Q_n	Q_{n+1}
0	0	0
0	1	0
1	0	1
1	1	1

Q_n is the output at a time before the 0–1 transition of the clock pulse. Q_{n+1} is the output after the next 0–1 transition of the clock pulse.

Figure 11.1.7 shows a useful circuit that latches a transient input and outputs a steady state at the first edge in. When the clock input goes 0–1, the Q output latches the D input. Further changes in the

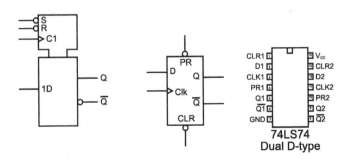

Figure 11.1.6 *The D-type latch*

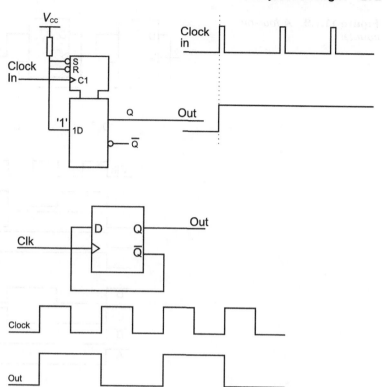

Figure 11.1.7 *An output latch*

Figure 11.1.8 *A divide-by-two circuit*

clock input will have no effect on the output. Only an input on the reset line will reset the Q output to 0.

When the $\overline{Q}$ output is connected to the D input, a divide-by-two circuit is achieved as shown in Figure 11.1.8. When Q = 0, $\overline{Q}$ = 1. At the next 0–1 transition of the clock, the output changes state as the 1 on the D input is output on the Q. So, Q = 1 and $\overline{Q}$ = 0. The input has to go through the 1–0 before the 0 on the D input is cascaded through to the Q output on the next 0–1 transition of the clock.

If you cascade four of these together, you will end up with a four-bit counter as shown in Figure 11.1.9. This is called an asynchronous or ripple-through counter since it takes the transition of the output of one stage to trigger the next. The downside of this is that there will be nanoseconds of time between one output number and the next where an incorrect number is output. This is illustrated in Figure 11.1.10. Whether or not this is important depends on the application.

If it is important that the correct sequence is maintained, a synchronous counter is needed (see Section 11.3).

Truncated counts

Sometimes, as in the Die problem (see question 2 of Problems 10.6.1), the full count of 0000–1111 is not needed. A die only counts from 1 to 6, and the original problem meant that the counter cycled through 0 to 5. In this circumstance, the D-type flip-flops could be reset when the count of 6 is achieved. That is, the counter would go through the sequence:

000–001–010–011–100–101–(110)–000

242 Higher Electronics

Figure 11.1.9 *A four-bit counter*

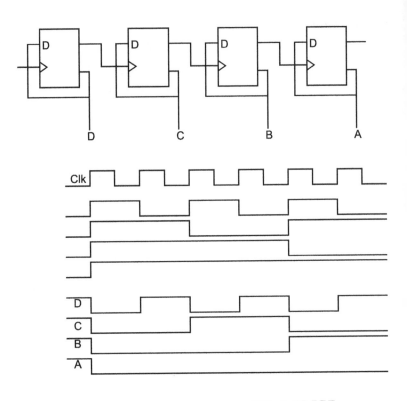

Ideal transitions 3 - 4
Actual outputs 3 - 2 - 0 - 4

3 0011 — 0100 4
2 0010 — 0000 0

Figure 11.1.10 *False counting states*

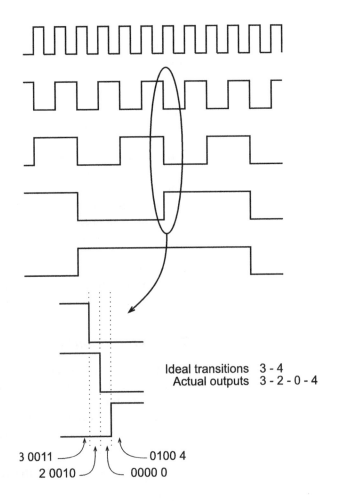

Figure 11.1.11 *Resetting a counter*

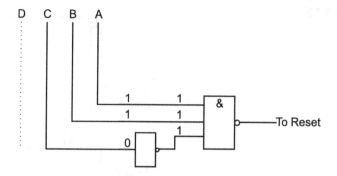

As soon as the count of 110 is detected, the counters would be reset. In a simple die circuit, the forbidden state ('output a 7') only lasts for nanoseconds, and so would not be seen.

A 74LS74 needs a 0 on the RESET line to reset the output, so the circuit of Figure 11.1.11 could be used to detect the 110 situation. In this situation, only a 3-bit counter is needed, the D being superfluous. The output of a NAND gate is only 0 when all of the inputs are 1. The circuit will only produce a 0 out when the code 110 is input. (Remember, in this circuit, A is the MSD – most significant digit.) As soon as the D-types reset, the output will be 000. The output of the NAND gate will be 1, so the reset condition will be released, and the D-types will be free to count once more.

Problems 11.1.1

(1) Use a 74LS74 D-type flip-flop to produce a circuit to give a continuous alarm after 12 input pulses.

(2) Design a counter using 74LS74 D-type flip-flops which will count through the sequence of 0–9 (0000–1001). This is the decimal counter.

(3) A counter circuit is fed from a 1 Hz clock. Design a circuit using 74LS74 D-types which:
 • continuously counts through 5 stable states (0–4); and
 • outputs a 1 s pulse each time the circuit is reset.

(4) A burglar alarm system consists of a microswitch on the door, an on–off keyswitch in the room and some digital electronics (Figure 11.1.12). Once set, the user has 7 seconds to leave the room before the alarm is active. On entering the room, the user has 7 seconds to disable the alarm before it sounds. A 2 Hz clock pulse is available. Design a suitable system.

(5) A motor controller circuit is operated by means of three momentary action push button switches marked 'Forwards', 'Reverse' and 'Stop'. To go forward, contacts 1A and 1B are made. To reverse, contacts 2A and 2B are made. Both relays should not operate simultaneously. Design a logic circuit using 4013 D-type flip-flops to operate the motor relays. The relay

244 Higher Electronics

Figure 11.1.12

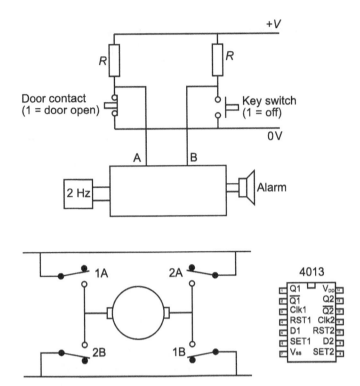

Figure 11.1.13

> contacts are shown in the unenergized state in Figure 11.1.13.
>
> (6) Each of two contestants in a quiz programme have a momentary push button to indicate that they know the answer to a question. The first one to push the button illuminates a light and locks out the other contestant. The quiz-master has an overall momentary push button to act as a reset. Design the logic circuitry needed.
>
> (7) An alarm system has:
> - one input;
> - two outputs: (a) horn; (b) indicator light;
> - two controls: (a) Accept; (b) Cancel.
>
> On receipt of an alarm signal (0–1), the horn sounds and the indicator lights up. Pressing 'Accept' shuts down the horn. Pressing 'Cancel' turns off the light and resets the system. It should not be possible to cancel before accepting. The circuit should latch any incoming alarm (so that subsequent 0–1–0 sequences do not cause a reset).

11.2 Digital clock generators

So far, we have looked at several ways of generating a square wave clock signal. One which has so far been missed is the *RC* oscillator based on logic inverter gates. Figure 11.2.1 shows an *RC* oscillator.

The frequency of oscillation is given approximately by:

$$f = \frac{1}{2.2 \times C1 \times R1}$$

*R*2 is usually very much bigger than *R*1 and is used to prevent the

Figure 11.2.1 *An RC oscillator using NOT gates*

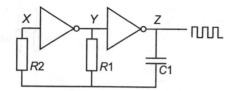

charge on $C1$ being discharged into the input protection diodes of the CMOS inverter. The oscillator works as follows.

When $X = 0$, then $Y = 1$ and $Z = 0$. $C1$ charges up through $R1$ and when $VC1$ passes the '1' input threshold of X, then the states change to $X = 1$, $Y = 0$ and $Z = 1$. $C1$ then discharges through $R1$ until the '0' threshold at the input of X is reached. The cycle starts again with $X = 0$, $Y = 1$ and $Z = 0$.

Example 11.2.1

Design a 1 kHz square wave oscillator.

$$1000 = \frac{1}{2.2 \times R1 \times C1}$$

Let $C1 = 0.1\,\mu\text{F}$. Then:

$$R1 = \frac{1}{2.2 \times 1000 \times 0.1 \times 10^{-6}} = 4.5\text{k}$$

Let $R2 = 47\text{k}$.

Problems 11.2.1

(1) Design an oscillator which produces a square wave output of approximately 500 kHz.
(2) Design an oscillator which produces a square wave output of approximately 1 Hz. (Note that the largest non-polarized capacitor could be as low as 2.2 µF.)
(3) Figure 11.2.2 shows the circuit diagram of a 'gated oscillator' where a control input inhibits or allows oscillations.
   (a) What input ('0' or '1') inhibits oscillation?
   (b) Design a 13 kHz oscillator.
(4) Design a 4 kHz gated oscillator using two NOR gates.

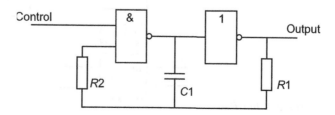

Figure 11.2.2

Figure 11.2.3 *The 4521 oscillator/frequency divider*

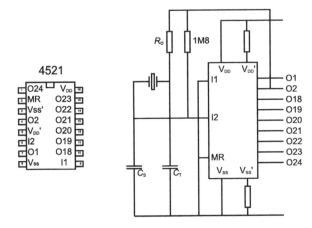

Clock generator circuits

As always, there is an integrated circuit designed to do the job, and in Figure 11.2.3 the 4521 is shown. This CMOS IC contains a chain of 24 toggle flip-flops with an overriding asynchronous master reset and an input circuit which allows for three different oscillator modes of operation (external frequency reference or *RC* oscillator or crystal oscillator). The application circuit shows it connected with a quartz crystal. A common crystal value seen in the catalogues is 4 194 304 Hz. This is because 2^{22} is 1 Hz. So, if this crystal were to be used in this circuit, then a 1 Hz clock could be obtained at O22− pin 14.

11.3 Synchronous design

J–K flip-flops

Before we can get far into synchronous design, we must first deal with yet another type of flip-flop. This is the J–K flip-flop. It has two inputs in addition to the clock pulse, called J and K. Among other things, the J input is a Set input and the K input is a Reset.

The truth table resolves the issue of:

- J = 0, K = 0, i.e. neither set nor reset so stay the same;
- J = 1, K = 1, i.e. both set and reset so change the state of the output.

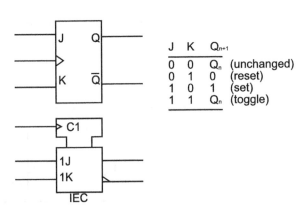

Figure 11.3.1 *The J–K flip-flop*

In most practical devices, there is also an asynchronous Set and Reset in exactly the same way as the 74LS74 dual D-type.

Master–slave vs. edge triggered operation

The D-type already met is an edge triggered circuit. This means that the active edge causes the defined operation to happen – that is, the logic level on the D input is transferred to the Q output. The two types of edge triggered inputs are shown in Figure 11.3.2.

The master–slave uses both edges of the clock input (Figure 11.3.3). The rising edge samples the input states while the falling edge implements the logic function and connects the output circuitry to the actual outputs. The IEC symbol which declared a master–slave output is shown next to the two outputs of the flip-flop.

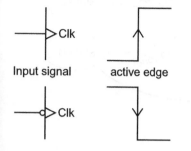

Figure 11.3.2 *An edge triggered input*

State diagrams

The clock inputs to synchronous circuits are usually all connected together. This is to avoid the ripple-through effect which was shown in asynchronous counter design in Section 11.1. The act of clocking the circuit can cause the design to change state, i.e. move from one stable set of outputs to another. The concept of a state machine is usually reserved for small logic circuits which perform a specific function and have a finite number of states. The circuit can change state according to:

- the current state;
- the inputs to the system;
- the current outputs of the system (although this is usually a function of the current state).

The state transition diagram is a diagram which shows all the current states and the inputs which cause a transition from one state to another. There are, in fact, two types of state machine:

- Moore machine – the outputs are dependent only on the present state. Simple counters fall into this category.
- Mealy machine – the outputs are dependent on the present state and the present inputs. This model allows for a far more sophisticated state machine.

In Figure 11.3.4 there are three inputs to the state machine and the movement between states is changed by the value of the three inputs 'start', 'X1' and 'reset'. It is not always necessary to place all of the inputs on the direction lines between each state. If an input is not described, it is assumed to be a 'don't-care' input.

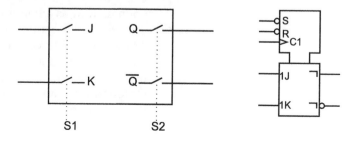

Figure 11.3.3 *A master–slave flip-flop*

Figure 11.3.4 *A state transition diagram*

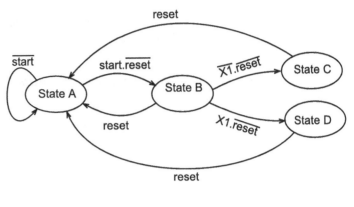

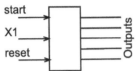

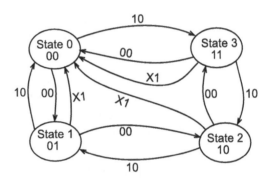

Figure 11.3.5

Example 11.3.1

Give the state transition diagram of a two-bit binary counter with UP/DOWN and RESET as the two inputs.

For compactness, inputs of UP and $\overline{\text{RESET}}$ would be indicated by 00. Note that the RESET function operates regardless of the state of the UP/DOWN control. Hence, this is written as an X – don't care.

Problems 11.3.1

(1) Draw the state transition diagram of a 3-bit (0–7) up counter with asynchronous reset.
(2) An electrically controlled garage door has two motor relays (up and down) and two control buttons 'operate' and 'stop' and two sensors 'fully-up' and 'fully-down'. Thus there are four states: stationary down, motoring up, stationary up and motoring down. Draw the state transition diagram of its operation.

Figure 11.3.6 *A 6-state counter*

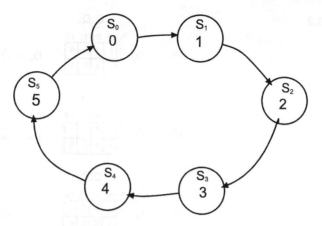

Synchronous counters

Figure 11.3.6 shows a simple 6-state up counter. This is going to be implemented in D-type and J–K flip-flops. Firstly, consider how each of these controls the change in the output state:

Current State Q_n	Next State Q_{n+1}	D input	J input	K input
0	0	0	0	X
0	1	1	1	X
1	0	0	X	1
1	1	1	X	0

The D input is straightforward, since it determines the next state. With the J–K flip-flop, the don't-care terms have to be used. In the first line, the Q output stays at 0. This could be an explicit reset $J = 0$, $K = 1$ or it could be a stay-the-same-transition $J = 0$, $K = 0$. This is why the K is noted as don't-care.

In this problem, a table can be created for the current-step/next-step transition. Firstly the D-type solution:

Current			Next					
QC	QB	QA	QC	QB	QA	DC	DB	DA
0	0	0	0	0	1	0	0	1
0	0	1	0	1	0	0	1	0
0	1	0	0	1	1	0	1	1
0	1	1	1	0	0	1	0	0
1	0	0	1	0	1	1	0	1
1	0	1	0	0	0	0	0	0

The next step is to translate the Q_n–Q_{n+1} transitions as a set of Karnaugh maps (Figure 11.3.7).

The expressions are somewhat complicated, but are fairly easily implemented (Figure 11.3.8).

Now the J–K flip-flop solution. The state table is more complex because there are the two variables J and K:

Figure 11.3.7

D_A Q_C

$Q_A Q_B$	00	01	11	10
0	1	1	0	0
1	1	X	X	0

$D_A = \overline{Q_A}$

D_B Q_C

$Q_A Q_B$	00	01	11	10
0	0	1	0	1
1	0	X	X	0

$D_B = \overline{Q_A}.Q_B + Q_A.\overline{Q_B}.\overline{Q_C}$

D_C Q_C

$Q_A Q_B$	00	01	11	10
0	0	0	1	0
1	1	X	X	0

$D_C = Q_A.Q_B + \overline{Q_A}.Q_C$

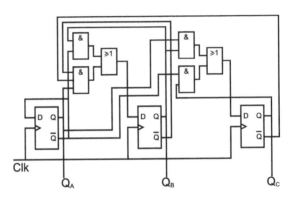

Figure 11.3.8 *A 0–5 counter*

Current			Next								
QC	QB	QA	QC	QB	QA	JC	JB	JA	KC	KB	KA
0	0	0	0	0	1	0	X	0	X	1	X
0	0	1	0	1	0	0	X	X	1	X	1
0	1	0	0	1	1	0	X	X	0	1	X
0	1	1	1	0	0	1	X	X	1	X	1
1	0	0	1	0	1	X	0	0	X	1	X
1	0	1	0	0	0	X	1	0	X	X	1

There will be six Karnaugh maps for this solution, but the resulting equations are simpler (Figure 11.3.9).

The solution for the synchronous counter becomes as shown in Figure 11.3.10.

Problems 11.3.2

(1) Design a synchronous 0–6 counter using:
   (a) D-type flip-flops;
   (b) J–K flip-flops.
(2) Design a synchronous counter with J–K flip-flops to output the number sequence:

      ... 3 4 5 6 7 8 9 3 ... etc.

Figure 11.3.9 *J–K Karnaugh maps*

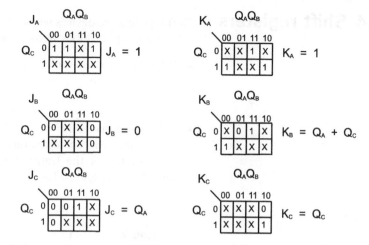

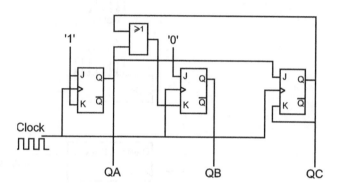

Figure 11.3.10 *A 0–5 counter*

(3) Design a synchronous counter with D-type flip-flops to count through the sequence:

... 0 1 2 5 6 7 0 ... etc.

(4) A logic circuit has two inputs, Clock and Reference, and four outputs, Q1, Q2, Q3 and Q4. Design the circuit so that the relationship between input and output is as shown in Figure 11.3.11.

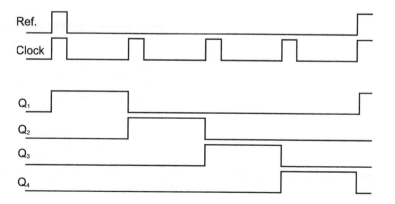

Figure 11.3.11

11.4 Shift registers

Shift registers are devices which shift a data stream according to the rate of an applied clock according to four register types:

- SIPO Data is presented to the register one bit at a time and advanced through the register at each clock pulse. The SIPO is used, amongst other things, as the receive stage of a microprocessor serial input port (Figure 11.4.2).
- PISO Data is loaded in parallel and data will advance through the register with each clock pulse. The PISO is used, among other applications, as the transmit stage of a microprocessor serial output port (Figure 11.4.3).

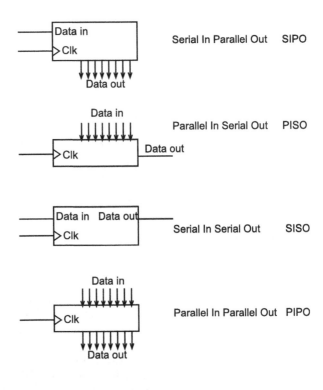

Figure 11.4.1 *Shift registers*

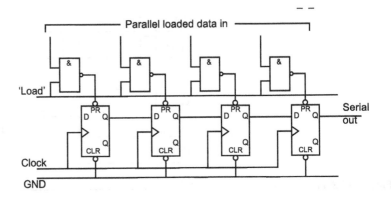

Figure 11.4.2 *Serial In Parallel Out*

Figure 11.4.3 *Parallel In Serial Out*

Figure 11.4.4 *Serial In Serial Out*

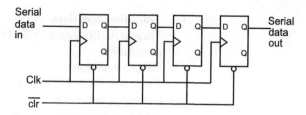

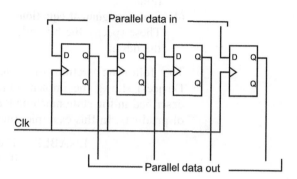

Figure 11.4.5 *Parallel In Parallel Out*

- SISO The SISO can be used as a delay circuit which is controlled by the clock rate (Figure 11.4.4).
- PIPO The PIPO loads data into the register at the clock pulse and it then becomes available at the outputs. More delays or register stores can be made by cascading several registers together (Figure 11.4.5).

Universal shift registers

The 74LS194 is a simple 4-bit bidirectional univeral shift register which can be parallel or serial loaded. The data can be available in parallel or serial (Figure 11.4.6).

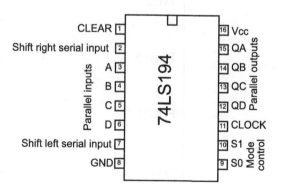

Figure 11.4.6 *The universal shift register*

11.5 Programmable logic

The output stage of the 16R8 PAL seen in Chapter 10 is an early and fairly unsophisticated device. The PAL design equations are necessarily more complicated in devices with flip-flops. In most software packages, the following would be required:

(1) **Transition equations**
These equations specify that the next state will be under various conditions
(2) **Output equations**
These are the outputs of the state machine.
(3) **Condition equations**
These equations specify the condition name for each set of input conditions which determine one or other of the transitions.
(4) **State-assignment equations**
These specify the bit code to be assigned to each state used in the design.

The following section of code (courtesy of Advanced Micro Devices) shows the implementation of a 3-bit up/down counter described in the state machine language used for the MACH series of products. In this example, there are two inputs:

> ENABLE '1' enables the counter
> '0' disables the counter

There are three outputs:

> CNT2 Most significant output
> CNT1
> CNT0 Least significant output

```
STATE
MOORE_MACHINE

ZERO        :=   UP      →  ONE
                 +DOWN   →  SEVEN
                 +STOP   →  ZERO

ONE         :=   UP      →  TWO
                 +DOWN   →  ZERO
                 +STOP   →  ONE

TWO         :=   UP      →  THREE
                 +DOWN   →  ONE
                 +STOP   →  TWO

THREE       :=   UP      →  FOUR
                 +DOWN   →  TWO
                 +STOP   →  THREE

FOUR        :=   UP      →  FIVE
                 +DOWN   →  THREE
                 +STOP   →  FOUR

FIVE        :=   UP      →  SIX
                 +DOWN   →  FOUR
                 +STOP   →  FIVE

SIX         :=   UP      →  SEVEN
                 +DOWN   →  FIVE
                 +STOP   →  SIX

SEVEN       :=   UP      →  ZERO
                 +DOWN   →  SIX
```

Sequential logic

		+STOP	→	SEVEN
ZERO.OUTF	=	$\overline{CNT2} * \overline{CNT1} * \overline{CNT0}$		
ONE.OUTF	=	$\overline{CNT2} * \overline{CNT1} * CNT0$		
TWO.OUTF	=	$\overline{CNT2} * CNT1 * \overline{CNT0}$		
THREE.OUTF	=	$\overline{CNT2} * CNT1 * CNT0$		
FOUR.OUTF	=	$CNT2 * \overline{CNT1} * \overline{CNT0}$		
FIVE.OUTF	=	$CNT2 * \overline{CNT1} * CNT0$		
SIX.OUTF	=	$CNT2 * CNT1 * \overline{CNT0}$		
SEVEN.OUTF	=	$CNT2 * CNT1 * CNT0$		
CONDITIONS				
UP	=	ENABLE * UP_DWN		
DOWN	=	ENABLE * $\overline{UP\_DWN}$		
STOP	=	$\overline{ENABLE}$		

In this code + means OR and * means AND. The first part shows how the eight states and their transitions are defined in terms of the CONDITIONS which in turn rely on the input logic pins.

The state output files such as ONE.OUTF = $\overline{CNT2} * \overline{CNT2} *$ CNT0 define how the outputs should be generated for each of the states (in this case ONE).

This code does not include the pin allocation sections as it is not necessary to get the flavour of how the state diagrams can be described to the software. To do more would require a full tutorial on the software package.

11.6 Memories

Memory is used for storing digital data. Memory is broadly divided into two categories:

(i) Random Access Memory (RAM) is a read–write memory which functionally can be further subdivided into two basic types:
  (a) static RAM;
  (b) dynamic RAM.
(ii) Read Only Memory (ROM) which is memory from which data can only be read.

Static RAM

Static RAM is basically a flip-flop which keeps its state as long as the power is applied. NMOS static RAM (whose use is now in decline) gave SRAM the reputation for high power consumption. The traditional CMOS structure achieved a very low operating current by giving an active function to the load, but the extra transistor structures needed make the die area large.

The advantages of SRAM are:

- **Byte wide**
  Small systems needing only a modest amount of any kind of memory could have their needs met by just one device.
- **No refresh**
  Large memory systems can afford the overhead of refresh circuits, but in small systems DRAM refresh would take up proportionally more cost, space and power.

- **High speed**
  CMOS SRAM can be optimized for speed.
- **Low operating power**
  Static CMOS circuitry consumes less power than DRAM devices.
- **Low standby power/data retention**
  Given a small cell to provide a backup supply, the SRAM will retain data for long periods.

Dynamic RAM

Dynamic RAM are constructed from cells with one transistor and one capacitor. The memory cells are arranged into a matrix of columns and rows. They are traditionally 1 bit wide, so sufficient RAM for (say) an 8-bit wide data system would require 8 DRAM devices. The cell capacitor is used to store the data, and a special sense amplifier is used to read the contents. The capacitor is very small and requires frequent (2 ms) refreshing if the stored data is not to be lost:

- **Addressing**
  To keep the size physically small, the DRAM uses a technique called multiplexing. For example, a 64k device should normally require 16 address pins; however, in a DRAM, there are only 8 address pins. In addition, there are two control lines: CAS (Column Address Strobe) and RAS (Row Address Strobe). Both are active low. In the case of the 64k RAM, the lower 8 bits of the address would be applied to the address pins, and RAS asserted. Then, the upper 8 bits of the address would be applied to the address bus and CAS asserted.
- **Refreshing**
  There are several acceptable methods of refresh, but the most popular (because, technically, it is the easiest) is the RAS Only refresh. This uses only the RAS strobe to refresh a specific row of the matrix. External circuitry must provide a row counter to keep track of the next row to be refreshed. Some microprocessors have this row counter built in.

The main advantages of Dynamic RAM are due to the cost per bit and the packing density for large memories.

Read Only Memories

EPROM

These are readily identifiable by the large quartz windows. The window itself is to allow the passage of ultraviolet light which is applied to erase the EPROM. In normal use (after programming), the window would be covered up with an opaque sticker to avoid accidental erasure.

EPROMs are available in CMOS and NMOS, and are large arrays of MOSFETS with a floating gate which is buried in the substrate. This floating gate is supplied uncharged (i.e. at a logic 1) and the process of programming the EPROM consists of converting the logic state of a MOSFET from a 1 to a 0. This is achieved by applying a 'high' voltage to the programming pin of the EPROM (12.5 or 21 V, depending on the type) and thus creating 'energetic' electrons which are able to 'tunnel' to the normally isolated floating gate. These memories store data by retaining a tiny charge (approx. 10^6 electrons) for a long time.

Manufacturers conservatively only guarantee charge storage for 10 years, but this is an extrapolation of high temperature worst-case conditions. In reality, they do not seem to lose any data unless the part is defective or damaged in some way. EPROMs have a limited endurance – i.e. there is a limit to the number of times which a part can be erased and reprogrammed. The data for this is once again uncertain, but most manufacturers will allow for up to 100 erase/reprogram cycles before degradation sets in.

The programming algorithm in the smaller (= older) parts meant that the elevated programming voltage had to be maintained for 50 ms. This has now been reduced for more recent/larger EPROMs (the 'smart' algorithm) and you now try a series of 1 ms pulses and test after each pulse to see if it stored satisfactorily. Once you are successful, you should apply an additional final write pulse equal to 3 times the previous total. Typically, most bytes programme first time, so you spend about 4 ms per byte.

Problem 11.6.1

Calculate the 'smart'-algorithm programming time for:

(a) a 64k × 8 EPROM;
(b) a 256k × 8 EPROM.

Mask ROM and fusible-link ROM

Mask ROMs are essentially custom chips manufactured with the desired bit-pattern built in. You must provide a semiconductor house ('silicon foundry') with the bit specification, and then it is converted into a custom metallization mask to hold the desired information. The setting up costs are not insignificant, and it is generally not economic unless you are producing in the thousands.

Many single chip microcontrollers include a few kilobytes of on-chip RAM and ROM so that a finished instrument need not have any external memory chips. However, most microcontroller families include versions that take external ROM and, in some cases, versions with on-chip EPROM.

EEPROM

These ROMs can be selectively erased and reprogrammed electrically while in circuit. They are useful for holding configuration information, calibration parameters, etc. that cannot be frozen before the computer or instrument is used. They use the same MOS floating-gate technique as EPROMs. They are much more expensive than EPROMs but can be reprogrammed many more times: manufacturers quote 100 000 read/write cycles at 25° operating temperature.

Although using a similar technology, there are also devices referred to as EAROMs where an individual byte can be reprogrammed.

12 Microprocessors

Summary

Microprocessors perform very simple tasks very quickly. It is this relative speed that gives them the reputation for 'intelligent' or 'smart' functions. They all have certain characteristics in common, and this chapter looks at these common functions before specializing in specific 'families' or groups of devices. There are dozens of types of microprocessor, but this chapter looks at one of the most popular industry standard parts – the Intel 8051.

12.1 Microprocessor fundamentals

Microprocessors fetch instruction codes from memory, find out what that instruction is commanding it to do, and then they do it. This is formally called the **fetch–execute cycle**. The three main areas of microprocessor-based systems are:

- **CPU** – central processing unit;
- **Memory** – read only or read/write;
- **I/O** – input/output.

Instructions can cause:

- data transfer between memory and the CPU;
- data transfer between the CPU and the input/output ports;
- data manipulation within the CPU.

Microprocessors can be called 4, 8, 16, 32 or even 64 bit machines. This characterization refers to the number of bits in a data word, i.e. the 'width' or size of data that is handled in one operation. (However, even this simple definition does not cover all situations: the Intel 8088 reads data in from memory in 8-bit chunks but deals with it internally in 16-bit words.) So when the data sheets refer to the 8-bit 8051, it means that instructions which cause data to be ADDED, SUBTRACTED, ANDED, etc. are to be performed 8 bits at a time – the 'word'. Another term for an 8-bit word is the 'byte'. Half of this is called a 4-bit 'nibble'.

Internally, all microprocessors have 'registers' which are capable of holding a dataword. The number of registers depends on the type of processor. Data in a register can be transferred to another register, a location in memory, or to an input/output device. One register which is common to all microprocessors is the **program counter**. This always holds the memory address of the next instruction to be executed. Another common register is the **instruction register**. This is where each instruction is held while it is being decoded. After that, the particular action to take place depends on the microprocessor. The **arithmetic and logic unit** is another common element in all microprocessors. This device allows data in registers or memory to be compared or subjected to various arithmetic/logical operations such as addition, subtraction, multiplication, division, logical 'and'ing, 'or'ing, 'not'ing or 'exclusive-or'ing.

12.2 The 8051 microprocessor

The 8051, like many microcontrollers, has the CPU, memory and I/O (input/output) integrated together in a flexible and extendable manner. It is an established product which is very popular in many different industries. The core features of the device are:

- 8-bit CPU optimized for control applications
- extensive Boolean processing (single-bit logic) capabilities
- 64K program memory address space
- 64K data memory address space
- 4 kbytes of on-chip program memory
- 128 bytes of on-chip data RAM
- 32 bidirectional and individually addressable I/O lines
- two 16-bit timer/counters
- full duplex UART
- 6-source/5-vector interrupt structure with two priority levels
- on-chip clock oscillator.

The basic architectural structure of the core is shown in Figure 12.2.1.

The MCS 51 microcontroller family comes in a variety of options. A selection of these are summarized in Table 12.2.1.

Table 12.2.1

Device name	EPROM (bytes)	ROM (bytes)	RAM (bytes)	8 bit I/O ports	Clock (MHz)	UART	ADC
8031	–	–	128	4	33	1	
8032	–	–	256	4	33	1	
8051	–	4k	128	4	33	1	
8052	–	8k	256	4	33	1	
8751	4k	–	128	4	33	1	
8752	8k	–	256	4	33	1	
80C452	–	–	256	5	16		
83C452	–	8k	256	5	16		
87C452	8k	–	256	5	16		
8 × C552	8k	–	256	5	24	$1 + I^2C$	8×8
8 × C652	8k	–	256	4	24	$1 + I^2C$	
87C752	2k	–	642	2 + 5/8	40	I^2C	5×8

The UART is a serial port used for communicating with other devices and microprocessors. The ADC is an analogue to digital

Microprocessors   261

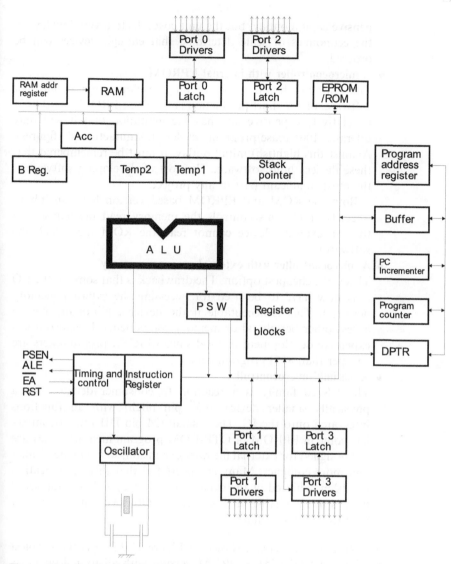

Figure 12.2.1 *The 8051 architecture*

converter which converts between an analogue voltage and a digital value.

Problem 12.2.1

One use for a microcontroller is as a washing machine controller. Suggest **three** other uses.

Basically, the options available from the MCS 51 family are:

- **A microcontroller with internal mask-programmed ROM**
  This is the option chosen for mass produced items. The act of specifying and creating a mask programmed ROM is an ex-

pensive capital outlay, but the unit cost falls to a very low level if the economies of scale determine that enough devices can be ordered.

- **A microcontroller with internal EPROM**
  This option is chosen for prototyping/developing microcontroller software, or for short production runs. The EPROM-based 8751 is relatively expensive and has the limitation that it will only tolerate 100 erase/program cycles (manufacturers' figures). Against this high(ish) initial outlay, it must be remembered that these devices hold software. The cost of developing software is the most significant part of any project.

  Both the ROM and EPROM based version have an inbuilt 'security fuse'. This controls the memory read mechanism, so that an external device cannot read the ROM (i.e. 'steal' the software).

- **A microcontroller with external program memory**
  This is the cheapest option. The drawback is that some of the I/O pins now have to be used for accessing the external memory device(s). This is acceptable if the device additionally has to access other peripheral or memory components. It is also a less expensive development method since EPROM programmers are cheaper than 8751 programmers.

- **A 'shrunk' microcontroller**
  The DS750 family is a reaction to consumer demand for a physically smaller device (0.3″ pin pitch) with all functions integrated onto the IC. These small (24 pin DIL) microcontrollers come in EPROM or OTP-ROM program memory. They are not designed for external memory expansion, so are made smaller and more compact. Many engineers feel that this simplification has been achieved at the expense of one major flaw – the serial port is not available in this package. Otherwise, it is generally held to be a good thing.

The 40 pin DIL version shown in Figure 12.2.2(a) is the simplest version, an 8051/8751 (EP)ROM version with all four 8-bit ports being used as I/O lines. Active-low lines are identified with a preceding oblique stroke (e.g. $\overline{EA}$). ('Active-low' means that a logic 0 is needed for the input to perform its stated function. In this case $\overline{EA}$ is pronounced 'not-External-Access' or 'not-E-A'. The implication is that in order to access external memory, this line should be tied low to 0 V.)

Common to all electronic circuits are the power supply lines V_{CC} and V_{SS}. In some variants, these may be called $V_{DD}(= V_{CC})$ or $GND(= V_{SS})$. Technically, this is a mix of terminology since V_{CC} is usually used for $+V$ bipolar transistor supplies and V_{SS} is usually used for MOS ground supplies. However, it is the terminology used by Intel. The letter 'C' in the designation implies a CMOS (low power variant). The power supplies needed are:

Voltage $V_{CC} = +5\,\text{V}$ $V_{SS} = 0\,\text{V}$
Current 160 mA/20 mA (8031/80C31) or
         250 mA/40 mA (8751/87C51)

The following section will cover the housekeeping input lines RST, XTAL1, XTAL2 and Vpp|/EA. (/PROG|ALE and /PSEN

Figure 12.2.2 *8051 pinouts*

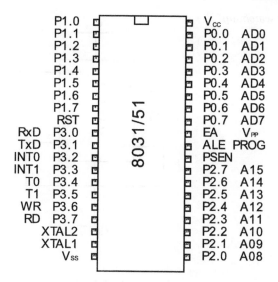

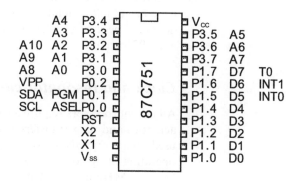

are outputs and will be explained later in the section 'Accessing external memory' (page 267).

RST is the system ReSeT line. It is essential to reset the microcontroller at power-up to initialize internal control registers. RST is an active high control line which will reset the microcontroller when a logic '1' is applied to it. The circuit of Figure 12.2.3 shows the basic configuration, with a switch added as an option to provide manual (additional) reset if required. The *RC* circuit provides a time delay which must hold the RST line HIGH for a minimum of 24 oscillator periods.

The watchdog timer

In some designs it is considered necessary to add a **watchdog** controller. This is a separate integrated circuit which will reset the microprocessor automatically if it is not regularly 'serviced' (sent a logic signal from the microcontroller to reset the watchdog). Since it is difficult to guarantee the integrity of any software, if an error does occur such as the microcontroller hanging up while waiting for an input which never arrives or getting stuck in a software loop, then the watchdog would not be serviced, so it would generate a reset.

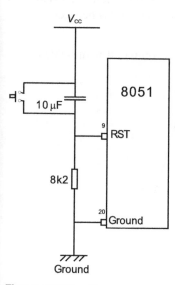

Figure 12.2.3 *The power-on reset*

Figure 12.2.4 *The watchdog timer*

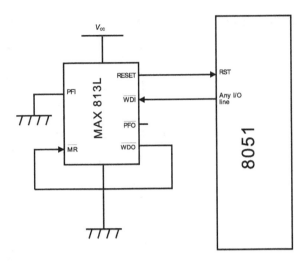

Figure 12.2.4 shows one type of watchdog controller that integrates the reset function. Some members of the 8051 family have an inbuilt watchdog that can be programmed into the users' designs if necessary. These include the 542, 550, 552, 558, 562, 580 and 592 variants.

Clock and resonator circuits

All microprocessors require a **clock**. It is used to synchronize all the activities and data transfers within the CPU. Originally, the 8051 used quartz crystals up to 12 MHz, but later designs are capable of operating at up to 40 MHz. Figure 12.2.5 shows a circuit suitable for a 12 MHz crystal. This sounds quite a fast rate (40 Mhz has period of some 25 ns) but the Intel family of devices divide this rate by 12 internally to produce the machine cycle – the basic interval which handles single byte instructions. This brings the machine cycle rate down to 1 MHz (or up to 3.33 MHz for a 40 MHz clock).

If a low cost or minimum component count is of prime importance, then consider the use of 3-terminal ceramic resonators as discussed in Chapter 9 (Radio Frequency and other techniques). The disadvantages are:

- not as accurate as crystals (0.5% as compared with 0.002%);

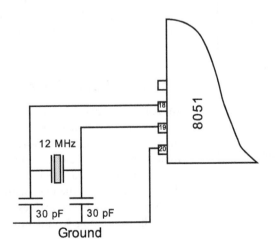

Figure 12.2.5 *The clock circuit*

Microprocessors **265**

- not as stable as crystals (as compared with 0.005%);
- not available in as high a frequency range as crystals (only up to 8 MHz).

Vpp/EA is a control pin which has two functions. V_{PP} is a programming voltage for the EPROM version. A high voltage (12.5 V or 21 V, depending on the version) is used to store the users' programs on the internal EPROM. An 8751 programmer will apply address and data signals to the device, and a controlled pulse applied to the ALE pin will use the high voltage on the Vpp pin to store the data.

/EA is a signal which informs the MCS 51 to look for its program in external memory. If /EA is high (+5 V), the device looks for its program in internal EPROM. If /EA is low (0 V) the device looks for its program in external EPROM. Several examples of applications of /EA will be given in the section on memory decoding.

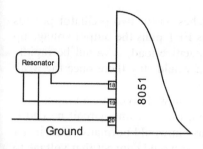

Figure 12.2.6 *The resonator circuit*

I/O port structures

All four ports in the MCS 51 family are bidirectional. Each consists of a latch, an output driver and an input buffer. When using the 8051/8751, all four ports can be used as inputs or outputs. Each I/O line can be independently used as an input or an output (although as an output, the pins can sink more than they can source).

Ports 1, 2 and 3 are called quasi-bidirectional. After the 8051 has been powered up or after a RESET, all ports act as inputs. The exception to this is when the $\overline{EA}$ (external access) pin is connected to 0 V in order to instruct the IC to access external program memory. In this case, ports 0 and 2 have special functions. The quasi bidirectional structures are shown in Figure 12.2.7. Ports 1 and 3 (and Port 2 if not used to access memory) have a 'weak internal pull-up' formed from a FET transistor switch. When used as an output, each pin of each of these ports will turn on the lower FET to output a 0 or it will turn off the lower FET to output a 1 via the pull-up resistor. This is the reason why the 8051 ports can sink more current than they can source. The lower FET represents a low impedance connection to ground when it is turned on, while the pull-up resistor (FET) is a higher impedance. The diagrams are actually a little oversimplified. In parallel with the pull-up 'resistor',

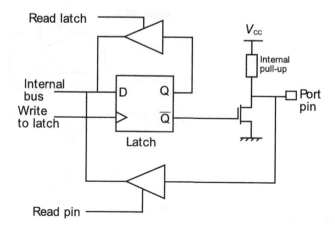

Figure 12.2.7 *Quasi-bidirectional port structure*

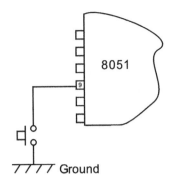

Figure 12.2.8 *A single push-button input*

there is another FET which switches on for two oscillator periods whenever a 1 is to be output. This FET pulls the output voltage up sharply. That is, if there is a capacitive load, it would be rapidly charged to the V_{CC} potential and would stay there once the weak pull-up was the only output device.

This facility is useful, because if you wish to create an input from (say) a push-to-make push-button, then the only circuit needed is that of Figure 12.2.7. Reading this pin would normally result in a 1 being input since the pull-up resistor would connect that voltage to the CPU buffer. Pressing the button would connect the pin to ground and hence a 0 V would be input.

When using the pin as an output, the only care that must be taken is that you must not write a 0 out on that pin before trying to use it again as an input. To do so would cause a conflict with any devices which may be connected to the pin. If an external device is trying to input a 1 and the 8051 is simultaneously trying to output a 0, then catastrophe will be the only result! The output FET of the 8051 is capable of sinking up to 15 mA which could damage sensitive devices which are trying to output a logic 1. Alternatively, if the external device trying to put a signal into the 8051 is capable of outputting more than 15 mA, then there is a good chance of damaging the input buffer of the 8051. In practice, though, it is rare to want to use a port as both input and output, so this situation should not arise.

Port 0 is different. This port has no weak pull-up because of its capability to read from and write to external memory devices. It has two output transistors which provide high current paths to either the $+V_{CC}$ or GND supplies. The pull-up FET only operates when executing an instruction which causes it to access external memory. The output pins for normal operation are **open-drain**, i.e. only connected to the lower pull-down FET. It would need external pull-up resistors to function in the same way as Figure 12.2.9. It is a true bidirectional port because when used as an input, the pins float (are not connected to any particular voltage).

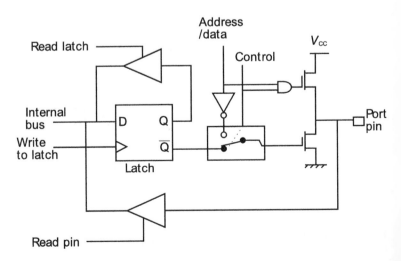

Figure 12.2.9 *Port 0 circuit*

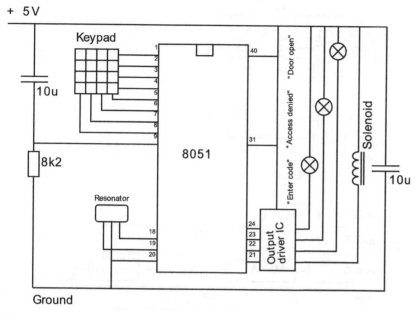

Figure 12.2.10 *A design example*

Design example

Figure 12.2.10 shows a simple example of an 8751 being used as a combinational door lock controller. Port 1 is used to operate a 16 key matrix keypad, while Port 2 operates the indicator lights and door solenoid.

The keypad is arranged in columns and rows. Pressing a key causes a column to be connected to a row. The MCS51 sequentially 'strobes' the columns and senses (via the rows) whether a key has been pressed. To save output lines, a dedicated IC (74C922) could be used to operate the keypad. In this way, the MCS51 needs to dedicate only 5 I/O lines.

> **Problems 12.2.2**
>
> (1) Design the hardware of two electronic dice (2 die) based in an 8751 processor. The outputs should be the conventional 7 'dots' (LEDs) while the inputs should be a 1–2 die selector switch and a 'throw' button.

12.3 Accessing external memory

If the /EA line of the microcontroller is LOW, then Port 0 and Port 2 are used to access the program memory. Every time a byte of data is to be fetched from this external memory the following sequence occurs:

- Port 0 outputs the low byte of the external memory address while Port 2 outputs the high byte of the external memory address.
- The ALE line (Pin 30) is asserted. This will cause the lower byte

Figure 12.3.1 *An output latch*

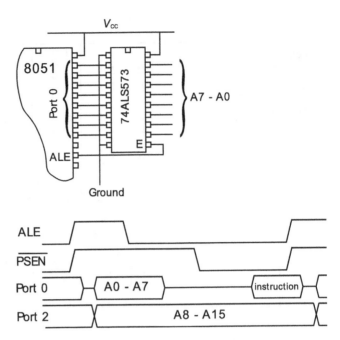

Figure 12.3.2

of the address to be stored by an external latch (see Figure 12.3.1).
- Port 0 is 'floated' (put in a high impedance mode) and then ...
- ... /PSEN (Pin 29) is asserted. This signal operates the external memory device (i.e. /CE of the EPROM).
- The byte of data is then read in by Port 0.

This is somewhat confusing as a word description, so Figure 12.3.2 shows the timing of the control signals. The sequence of these signals (ALE and /PSEN) is started automatically by the action of setting /EA low. Additional circuitry would be needed to use these signals and Figure 12.3.1 shows how this could be achieved. The latch could be a 74LS373 or 74LS573 octal latch. These devices store a byte of data when the E (enable) line is activated.

Logical separation of program memory (read only) and data memory (read/write)

An original design feature of the 8051 is that it has separate program and data memory. The microcontroller uses /PSEN (pin 29) to access program memory and /RD (Pin 17) and /WR (Pin 16) to access data memory. The logical separation of these two memory spaces allows the data memory to be accessed by 8-bit addresses which can be more quickly stored and manipulated by an 8-bit CPU (although 16-bit addresses can be used if necessary).

Figure 12.3.3 shows the timing of data transfers, while Figure 12.3.4 shows how the 8031 could support both program memory in EPROM and data memory in RAM. The distinction is due to the fact that:

- the /PSEN signal is used to access the EPROM;
- the /RD–/WR are used to access the RAM.

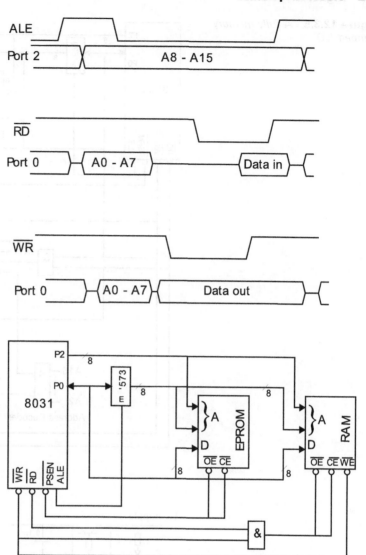

Figure 12.3.3 *Timing of data transfers*

Figure 12.3.4 *How the 8031 can support program memory in EPROM and data memory in RAM*

Larger size EPROM and RAM devices can be used directly up to 64 kbyte each. Designers of embedded systems are frequently as interested in target board size as much as cost, so the usual solution is to use as large a memory device as possible, i.e. use 1×64 kbyte instead of 4×16 kbyte. However, Figure 12.3.5 shows a fully decoded example with segmented memory and memory mapped I/O.

Non-volatile memory

There are many situations where it is important not to lose the contents of memory when the power is removed. The two main solutions for this use EEPROM or battery/capacitor backed RAM.

EEPROM are available in serial or parallel access. Serial EEPROM need special algorithms to save and read data. Parallel EEPROMs are easier to drive since they more closely resemble conventional EPROMs.

Figure 12.3.5 *A fully memory mapped I/O*

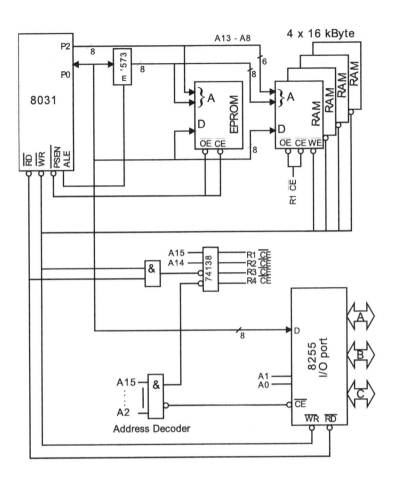

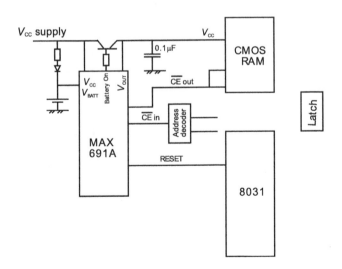

Figure 12.3.6 *Controlling non-volatile memory*

RAM whose contents are preserved with backup capacitor or battery are most conveniently interfaced via proprietary control ICs. The difficulty with RAM is that if is there is any noise or jitter on the /WE and /CE lines as the main system power shuts down, then parts of memory will be lost. Several manufacturers produce suitable ICs which can be used as in Figure 12.3.6.

Alternate use of Port 3

Port 3 and two lines of Port 1 have alternate functions. These can be used selectively and have the following use:

PORT PIN   Alternate function
* P1.0   T2   (Timer/counter 2 external input)
* P1.1   T2EX   (Timer/counter 2 capture/reload trigger)
  P3.0   RXD   (serial input port)
  P3.1   TXD   (serial output port)
  P3.2   /INT0   (external interrupt 0)
  P3.3   /INT1   (external interrupt 1)
  P3.4   T0   (Timer/counter 0 external input)
  P3.5   T1   (Timer/counter 1 external input)
  P3.6   /WR   (external data write strobe when /EA is low)
  P3.7   /RD   (external data read strobe when /EA is low)

(* P1.0 and P1.1 serve these alternate functions only on the 8052 and more complex MCS family parts which have more than just the two timers.)

Intel use the dot notation of the form P3.4 to indicate bit 4 of Port 3.

12.4 Internal architecture

We cannot make much more forward progress with the device until more is revealed about the internal architecture of the 8051. Figure 12.2.1 showed the internal structure with the four ports, two timers, internal registers and all the other bits and pieces which make up the microcontroller. The base model has 128 bytes of available RAM addressed from 00 to 7F (hexadecimal notation – see Appendix A for an explanation of this useful numbering system).

There is an additional area of RAM from 80 to FF called the **special function register** (SFR) space which holds registers which control or report on the many functions built into the IC. These two memory areas are called the **direct memory space**. All of these addresses can be read or written to, but, obviously, the SFR space is reserved for the special functions. Not all of the 127 locations in SFR are used.

Registers

There are eight general purpose registers, called R0, R1, R2, R3, R4, R5, R6 and R7. They are not physical registers in the CPU as such. They actually occupy general purpose RAM addresses 00–07. The 8051 allows the programmer to have four sets of these eight registers. These are called Register Banks 0, 1, 2 and 3, but at any one time there are only eight active registers called R0–R7. At power up, the machine defaults to Register Bank 0.

Register Bank	Address in RAM of R0 .. R7
0	00–07
1	08–0F
2	10–17
3	18–1F

So, if Register Bank 0 is selected, storing a number in Register R3 would cause it to be saved at internal RAM location 03. But if Register Bank 1 is selected, storing a number in Register R3 would cause it to be saved at internal RAM location 0B. Obviously, there is a potential problem here which could be caused by moving data from a register when the wrong Register Bank is selected. The wrong data would be moved. This is the responsibility of the programmer.

As far as the special function register is concerned, the following list shows the various functions of the direct bytes:

Direct byte address	Hardware register symbol	Function
F8H		
F0H	B	B Register
E8H		
E0H	A or ACC	A Register
D8H		
D0H	PSW	Program Status Word
C8H		
C0H		
B8H	IP	Interrupt Priority
B0H	P3	Port 3
A8H	IE	Interrupt Enable
A0H	P2	Port 2
99H	SBUF	Serial Data Buffer
98H	SCON	Serial Control
90H	P1	Port 1
8DH	TH1	Timer 1 High Byte
8CH	TH0	Timer 0 High Byte
8BH	TL1	Timer 1 Low Byte
8AH	TL0	Timer 0 Low Byte
89H	TMOD	Timer Mode
88H	TCON	Timer Control
83H	DPH	Data Pointer High Byte
82H	DPL	Data Pointer Low Byte
81H	SP	Stack Pointer
80H	P0	Port 0

Microprocessor instructions

The **hardware register symbol** is the term for each register coined by Intel and it is also the term which is accepted by all assembler software (which converts the 8051 instructions into machine code). The address is just as acceptable a term as the symbol, so instructing the CPU to write (move) the hexadecimal number 3A to address 80 would be achieved with the instruction:

mov 80H, #3AH

More readable would be the instruction:

mov P0, #3AH

In both cases, the binary number 00111010 would be written out on Port 0. Most assemblers use the suffix H to indicate hexadecimal numbers, B for binary numbers (and sometimes O for octal numbers). However, there are a few which use a H (or B or O) prefix with the number itself in single quotes, e.g. H'3A'. Sometimes when handling control registers, it helps to fix the bit pattern more visually if binary is used, as in:

mov P0, #00111010B

The # symbol is used by Intel to indicate that it is the number itself which is to be loaded. The difference would be:

```
mov P0, #00111010B;  output 00111010 to Port 0
mov P0, 00111010B;   output the contents of
                   ; location 00111010 (3A) to Port 0
```

The full instruction set is shown in Appendix B.

Note that any register in the SFR which has an address ending in −8 or −0 is bit addressable. This means that the dot notation can be used to read or write to an individual bit in that register. The following are valid instructions:

```
      setb P0.4;  Setting one bit of output port P0
       clr IE.0;  Clearing bit 0 of
                ; the Interrupt Enable Register
   jb a.0, next; jump if bit 0 of the Accumulator = 1
                ; to program label 'next'
```

If the register is not bit addressable, individual bits can be manipulated with the logical OR or AND instructions, for example:

```
orl TMOD, #01000000B;  set bit 6
                     ; orl = logical OR
anl TMOD, #10111111B;  reset bit 6
                     ; anl = logical AND
                     ; take care where and how you
                     ; pronounce them
```

Appendix A explains more about hexadecimal numbers and arithmetical/logical manipulation of this sort.

The other registers are all special function and are:

- **A register or accumulator**

  This register has its counterpart in most 8-bit microprocessors. It is connected to the output of the ALU (Arithmetic and Logic Unit) and is the repository for the result of most arithmetic and logic operations. It is also the source of data for many instructions.

- **B register**

  This is mostly used in multiply and divide operations. If not required for these functions, the memory space is available as a general purpose 8-bit register.

- **PSW (Program Status Word)**

  The status register exists in one form or other for all microprocessors. It reports on the nature of the result of some arithmetic and logic operations. It reports the state of the A register. It allows the programmer to select which of the register banks is in use:

PSW.7	PSW.6	PSW.5	PSW.4	PSW.3	PSW.2	PSW.1	PSW.0
CY	AC	F0	RS1	RS0	0V	–	P

CY PSW.7 Carry Flag
            Set/cleared by hardware or software
            during certain arithmetic and logical instructions

AC PSW.6 Auxiliary Carry Flag Set/cleared by hardware during addition or subtraction instructions to indicate carry or borrow out of bit 3

F0 PSW.5 Flag0
Set/cleared/tested by software as a user-defined status flag

RS1 PSW.4 Register Bank select bit 1

RS0 PSW.3 Register Bank select bit 0

0V PSW.2 Overflow flag Set/cleared by hardware during arithmetic instructions to indicate overflow conditions

– PSW.1 (reserved)

P PSW.0 Parity Flag
Set/cleared by hardware each instruction cycle to indicate an odd/even number of 'one' bits in the accumulator

The working register bank would be selected according to:

RS1 PSW.4	RS0 PSW.3	Register Bank selected	Internal RAM address
0	0	0	00–07
0	1	1	08–0F
1	0	2	10–17
1	1	3	18–1F

One way to select a register bank would be to write directly to the PSW:

    mov psw, #00010000B; Select Register Bank 2

This is acceptable, but it does override some of the status bits. A better way would be to alter the appropriate PSW bits directly as in:

    setb psw.4; Set bit 4
    clr psw.3; Clear bit 3

or even by:

    anl psw, #11100111B; Select Register Bank 0
    ; (clear bits 4 and 3 first)
    orl psw, #00010000B; Select Register Bank 2

Most assemblers will even recognize the abbreviations RS1 and RS0, allowing:

    setb   rs1
    clr   rs0

which is even more meaningful.

Some instructions test bits in the PSW:

    jc next; test the overflow bit (carry)
    ; and jump if it is set to label 'next'

IP Interrupt Priority This sets the priority of the five interrupt sources

P3 Port 3 A general purpose input/output port. It also has alternate functions which allows it to act as I/O for serial, timer and external interrupt signals

IE	Interrupt Enable register	
P2	Port 2	Also used to provide the upper byte of external memory address
SCON	Serial Port Control/Status Register.	
P1	Port 1	
TCON	Timer/Counter Control/Status Register	
SP	Stack Pointer	
P0	Port 0	

The following (trivial) program shows how the input from Port 3 could be copied onto Port 1:

```
          org   0
    loop  mov   A, P3
          mov   P1, A
          jmp   loop
          end
```

This shows the assembler directives **org** and **end**:

org tells the assembler to assemble code from memory location 0000. This is where the CPU looks for its first instruction after a RESET

end instructs the assembler to stop assembling

Problems 12.4.1

(1) Write a program to output the contents of Port 0 onto Port 1 and the complement of Port 0 onto Port 2.
(2) Write a program to input a number on Port 0, input another number on Port 1, add them together and output the result on Port 2.

Timer/counters

The 8051 has two 16-bit timer/counter registers – Timer 0 and Timer 1. When activated, these registers increment once for each event. An event can be related to the system clock (timer) or from an external logic signal transition (counter). If required, a timer/counter register can be set up to generate an interrupt when it 'rolls over' from all 1s to all 0s, i.e. when it reaches its full possible count of FFFFH (65 535) and advances the register to 0000 on its next increment. This timer/counter register is called TL0 and TH0 for Timer 0 or TL1 and TH1 for Timer 1. The 'L' and 'H' refer to the low and high byte of the 16-bit register respectively.

The timer/counter register can be preset to any value so that any time or count limit can be set up. For example, to time or count 20 events, the register should be loaded with the number FFDCH (65 616). There are several modes of operation of the timer/

Figure 12.4.1

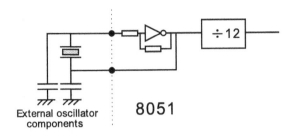

External oscillator components

counters. The hardware inputs for the timer/counter channels 0 and 1 are port pins P3.4 and P3.5 respectively.

When used as a timer, deriving its count from the system clock, it first divides that clock by 12 as in Figure 12.4.1.

The operation of Timer 0 and Timer 1 is effected by:

    Timer Mode Control Register   TMOD   (address 89)
    Timer Control Register   TCON   (address 88)

For the Timer Mode Control Register:

TIMER 1	TMOD.7	GATE	When gate = 1, Timer 1 is enabled only when INT1 = 1 and TR1 = 1
	TMOD.6	C/T	Timer/counter select: 0 = Timer 1 = Counter
	TMOD.5	M1	mode
	TMOD.4	M0	mode
TIMER 0	TMOD.3	GATE	When gate = 1, Timer 0 is enabled only when INT0 = 1 and TR0 = 1
	TMOD.2	C/T	Timer/counter select: 0 = Timer 1 = Counter
	TMOD.1	M1	mode
	TMOD.0	M0	mode

M1	M0		Operating Mode
0	0	mode 0	13-bit timer/counter
0	1	mode 1	16-bit timer/counter
1	0	mode 2	8-bit auto reload timer/counter
1	1	mode 3	timer/counter stopped

For the Timer Control Register:

TCON.7	TF1	Timer 1 overflow flag. Set when timer/counter overflows. Cleared by hardware when the processor vectors to the interrupt routine.
TCON.6	TR1	Timer 1 run control. Set or cleared by software to turn the timer/counter on or off
TCON.5	TF0	Timer 0 overflow flag. Set when timer/counter overflows. Cleared by hardware when the processor vectors to the interrupt routine.
TCON.4	TR0	Timer 0 run control. Set or cleared by software to turn the timer/counter on or off
TCON.3	IE1	Interrupt 1 edge flag. Set by hardware whenever an external interrupt is detected. Cleared when the interrupt is processed.

TCON.2	IT1	Interrupt 1 Type control bit: 0 = low level input trigger
		1 = falling edge input trigger
TCON.1	IE0	Interrupt 0 edge flag. Set by hardware whenever an external interrupt is detected. Cleared when the interrupt is processed.
TCON.0	IT0	Interrupt 0 Type control bit: 0 = low level input trigger
		1 = falling edge input trigger

- **Mode 0**

  This is a (virtually obsolete) mode which is included for functional compatibility with the earlier 8048 microprocessor family. In it, both Timer 0 and Timer 1 are 13-bit timer/counters with a divide-by-32 prescaler.

- **Mode 1**

  This is the same as Mode 0, except that the Timer register is run with all 16 bits of TL0 and TH0 (or TL1 and TH1).

$$\text{mov TMOD, \#10010001B}$$

In this example, both channels are set up to be timers in Mode 1. Timer 1 is to be controlled by the external pin 13, P3.3 while Timer 0 is to be controlled by the software setting of TR0. In Mode 1, both of the timers are 16-bit and so TH and TL must be loaded. Timer 1 is to be loaded with F000 and Timer 0 with D000.

```
mov TH1, #F0H;  Timer 1 register initialization
mov TL1, #00H
mov TH0, #D0H;  Timer 0 register initialization
mov TL0, #00H
```

The TCON timer control needs setting up, but in this example it is not proposed to use interrupts, so it becomes a simple task:

$$\text{mov TCON, \#00000000B}$$

Thereafter, Timer 1 can be started with a signal on External Interrupt 1 while Timer 0 can be started with an instruction such as:

$$\text{setb TR0}$$

For polling purposes, each byte of the timer/counter registers would have to be monitored. The following code locks up progress until both high and low bytes of Timer 0 advance to zero:

```
wait1: cjne TH0, #00, wait1;  if TH0 is not = 0
                            ;  jump to wait1
       cjne TL0, #00, wait1;  if TL0 is not = 0
                            ;  jump to wait1
```

The first line compares the contents of timer register TH0 with 00 and jumps if it is not equal (to 00) to the line with the label wait1. The second line compares the contents of the timer register TL0 with 00 and jumps if it is not equal (to 00) to the line with the label wait1. The overall effect is that the code will 'hang up' on these two lines until the 16-bit register TH0 and TL0 advances from its initial setting of D000 to 0000. Polling is cumbersome and wasteful of time, so all but the most trivial programmes tend to use interrupts. It might be risky for the externally controlled Timer 1 to hang up in such an endless loop in case the external signal was lost or changed unexpectedly.

- **Mode 2**

  This useful mode configures the timer register as an 8-bit counter with automatic reload. When TL0 (or TL1) overflows, it sets the Timer Flag TF0 (TF1) and reloads TL0 (TL1) with the contents of TH0 (TH1). The reload leaves TH0 (TH1) unchanged. This mode is quite often used for baud rate generation for the serial communication port, or for any application which requires a regular frequency to be output. When setting up, load both TH0 and TL0 (TH1 and TL1) with the reload figure since the TL register is only reloaded when it overflows. Otherwise, the first timing run could be slightly longer if the TL register is at its reset value of 00.

- **Mode 3**

  This mode affects the two timers in different ways.

  Timer 1 in Mode 3 simply stops counting. The effect is the same as setting TR1 = 0.

  Timer 0 in Mode 3 splits the TL0 and TH0 registers into two independent 8-bit counter timers. TL0 is controlled in exactly the same way as in mode 1, i.e using control bits C/T GATE and TR0. TH0 can only act as a timer and uses the Timer 1 control functions TR1 and TF1. Timer 1 loses its ability to generate interrupts if Timer 0 is set into Mode 3. However, this mode finds a use in the circumstances when an extra timer function is required.

The following example shows how to generate a 50 Hz square wave on port 1 bit 0. 50 Hz requires a 10 ms on and 10 ms off cycle which can be derived from a 12 MHz clock if 10000 machine cycles are counted for each ON and each OFF period. To make a timer count up to 10 000, the 16-bit register must actually be loaded with 65 536 − 10 000 = 55 536 − which is D8F0H:

```
; 50 Hz clock from a 12 Mhz crystal
org 0

        mov     TH0, #D8H
        mov     TL0, #F0H           ; set up register counts
        mov     TCON, #0            ; no interrupts
        mov     TMOD, #00000001B    ; timer 0 in mode 1
        setb    TR0                 ; start timer 0
label:
        cjne    TL0, #00, label     ; wait until low byte overflows
        cjne    TH0, #00, label     ; wait until high byte overflows

        mov     TH0, #D8H           ; reset counter high byte
        mov     TL0, #F0H           ; reset counter low byte
        cpl     P1.0                ; change state of the output pin

        jmp     label               ; loop to do it all again

        end
```

Interrupts

A CPU executes a program instruction by instruction. It will work its way through the instructions sequentially unless one of two events occur:

(i) The current instruction is an explicit command to jump a set number of instructions or to jump to a particular address.
(ii) An interrupt occurs. If interrupts have been enabled then that interrupt will cause the CPU to jump to a fixed location.

The 8051 has a useful vectored interrupt system that allows for five interrupt sources which are (in the order of priority):

Interrupt	Flag name	Vector
External Interrupt 0	IE0	0003
Timer Counter 0	TF0	000B
External Interrupt 1	IE1	0013
Timer Counter 1	TF1	001B
Serial Port	RI + TI	0023

The 'vector' address represents the address to which the CPU will relocate in the event of the named interrupt occurring. What will actually happen is that the program counter register holds the address of the next instruction to be executed. When the interrupt occurs, the program counter saves that next address on the memory area known as the stack and loads one of the above addresses into the program counter instead. If, for example, IE1 (external interrupt 1) had been enabled and at a particular point in the program, Port 3 bit 3 (pin 13) was taken from a logic 1 to a logic 0, then an interrupt would be generated. The program counter would be loaded with the address 0013 and after the current instruction had finished, then the CPU would fetch its next instruction from location 0013. Each interrupt has been allocated 8 bytes, and so if the interrupt is to take more code than this, then the simplest action to place a jmp (jump) instruction at address 0013.

The code which handles the interrupt is commonly called the **interrupt service routine**. The end of the interrupt service routine is terminated with the instruction reti (RETurn from Interrupt). This retrieves the return address from the stack and loads it into the program counter. Hence, the CPU resumes with the next instruction it was due to execute before the instruction occurred.

One last point. The vectors for the interrupt service routines are located fairly low in memory. If an interrupt does not happen, then after a power-up or system reset, the CPU will quite naturally execute code from locations:

0000   0001   0002   0003   0004   0005   etc.

It will quite happily work through the addresses, not knowing that the code at location 0003 contained code for External Interrupt 0. The way around this is to start the program something along the lines of:

```
org         0000
ajmp        start;          bypass vectors
```

```
                    org         0003H
                    ajmp        isr_ie0;        external interrupt 0
                    org         000BH
                    ajmp        isr_tf0;        timer 0 interrupt
                    org         0013H
                    ajmp        isr_ie1;        external interrupt 1
                    org         001BH
                    ajmp        isr_tf1;        timer 1 interrupt
                    org         0023H
                    ajmp        isr_ti_ri;      serial interrupt
            start:              ;               first piece of code
```

where, for example, jmp isr_tf1 represents an instruction to jump to the piece of code labelled isr_tf1.

Here is a simple program which is designed for an 8051 operating from a 12 MHz clock and merely outputs a slow square wave from Port 1.0:

```
start:  .equ        0000h
timer0: .equ        000Bh

        .org start
        jmp         main

        .org timer0                             ; timer vector
        jmp         timer_routine
main:
        mov         sp, #20h                    ; move stack to allow
                                                ; use of
                                                ; register bank 1
        setb        rs0                         ; register bank 1
        mov         r7, #0                      ; use r6 & r7 to count
                                                ; the number of
                                                ; interrupts
        mov         r6, #0
        clr         rs0                         ; register bank 0

        ;*** set up timer registers ***
        mov         tmod, #00000010b            ; auto reload TH to TL
        mov         th0, #6                     ; 250 counts @ 12MHz
                                                ; = 250 µS
        mov tl0, #6
        setb et0                                ; enable timer
                                                ; interrupts
        setb ea                                 ; enable interrupts
        setb tr0                                ; start count

        jmp         $                           ; hang up until
                                                ; interrupts occur
                                                ; jmp $ means jump to
                                                ; the current line

        ;*** interrupt service routine ***

timer_routine:
        setb        rs0                         ; select second register
                                                ; bank
        inc         r7
```

```
            cjne    r7, #40, timer_end  ; count 40 register
                                        ; R7 increments
            mov     r7, #0
            inc     r6
            cjne    r6, #100, timer_end ; count 100 register
                                        ; R6 increments
            mov     r6, #0

            cpl     p1.0                ; or whatever code is
                                        ; needed
                                        ; this toggles the output
                                        ; state
                                        ; of output port in
                                        ; P1.0 at
                                        ; 1 sec intervals
timer_end:
            clr     rs0
            reti

            .end
```

Some words must shortly be said about the nature of the 8051 stack, but first we must look at the interrupt control structure more closely. Special Function Register IP affects the priority of the interrupts. It was stated initially, that the priority of interrupts was:

$$IE0 \quad TF0 \quad IE1 \quad TF1 \quad TI+RI$$

This means that in the case where all interrupts have been enabled, then if the CPU is handling <external interrupt 1> and <timer 0> interrupt occurs, the CPU will:

- suspend execution of isr_ie1 code;
- deal with isr_tf0 code and when it has finished;
- return to where it was, in this case to isr_ie1 code.

However, since it has a lower priority, if a <timer 1> event occurs while the CPU is handling <external interrupt 0> then the code for isr_tf1 will just have to wait until the CPU has finished with isr_ie0. One way to alter the priority is use the IP register. This has five active bits which alter the priorities according to:

IE0	IP.0	External Interrupt 0
TF0	IP.1	Timer 0 Interrupt
IE1	IP.2	External Interrupt 1
TF1	IP.3	Timer 1 Interrupt
RI+TI	IP.4	Serial Port Interrupt

An instruction such as mov ip, #00001001B would set TF0 and RI+TI as high priority interrupts.

The priority order now becomes:

TF0
RI+TI
IE0
IE1
TF1

Figure 12.4.2 shows this diagrammatically.

Figure 12.4.2 *Interrupt priority*

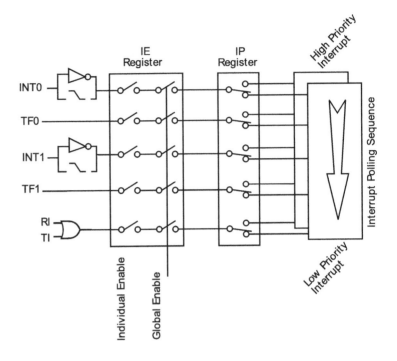

To enable interrupts, the Interrupt Enable IE register must be used:

EA	IE.7	Enable All control bit	0 = all interrupts disabled
	IE.6		
	IE.5		
ES	IE.4	Enable Serial port	1 = enable serial interrupts
ET1	IE.3	Enable Timer 1	1 = enable Timer 1 interrupts
EX1	IE.2	Enable External 1	1 = enable external 1 interrupts
ET0	IE.1	Enable Timer 0	1 = enable Timer 0 interrupts
EX0	IE.0	Enable External 0	1 = enable external 0 interrupts

```
mov   ie, #10001010B ; enable timer 0 and timer 1 interrupts
clr   ea             ; clear IE.7... disable all interrupts
```

Stack

At a reset, the stack pointer is set to 07. This is an 8-bit register which is used for holding the address of a position in memory for temporary storage of data and addresses. When data is pushed onto the stack, the stack is first incremented and then data is stored at that stack pointer address. A pop retrieves data from the current stack pointer address and then decrements the stack pointer. So, the first interrupt to occur would store the return address for CPU code at locations 08 and 09. Note that the default setting for the stack will cause it to overwrite the register bank 1. (Recall that Register Bank 1 sets R0–R7 as using RAM memory addresses 08–0F.) It does not overwrite the interrupt vector addresses because of the way in which the 8051 separates program code and data. Interrupt vectors would be part of the program code, while register banks are RAM.

The usual thing to do in any non-trivial application is to adjust the stack location, for example:

    mov sp, #30H ; relocating the stack to start at RAM adress 30

The stack grows as necessary. Every push instruction or subroutine call or interrupt causes the next address(es) in the stack memory to be used. The limit to the size of the stack is the available data RAM.

There is an interesting modification suggested by Philips Semiconductors which allows the 8051 to seem to have five external interrupts. EI0 and EI1 are already known about. Two others can be created by setting timer/counters 0 and 1 into 8-bit counter mode (Mode 2) with a preloaded count of FF. Thus, one more falling edge input on the Timer 0 or Timer 1 input would cause the counter to roll over and generate an interrupt. The interrupting counter would automatically be reloaded with FF from the TH0 or TH1 register so that the interrupt could occur again as soon as the interrupt service routing had been completed. To set up both timers for this type of operation:

    mov   TMOD, #01100110B ; both timers in mode 2
                           ; in counter mode
                           ; no gate control
    setb TR0   ; start timer 0
    setb TR1   ; start timer 1
    setb ET0   ; enable timer 0 interrupts
    setb ET1   ; enable timer 1 interrupts

Although covered in the following section, the serial Receive Data input line can also be used to generate an interrupt in the same manner. The serial port must be set into Mode 2 (a 9 bit UART with the baud rate determined by the oscillator). The setup would be:

$$SM0 = 1$$
$$SM1 = 0$$
$$SM2 = 0$$
$$REN = 1$$

With these parameters set up, the first 1 to 0 transition of the $R \times D$ pin would create an interrupt.

Serial communication

The 8051 family has a full duplex serial transmission circuit. The two pins P3.0 and P3.1 have the alternate function Receive Data ($R \times D$) and Transmit Data ($T \times D$).

Full Duplex means that it can transmit a byte of data through the $T \times D$ output at the same time as it is receiving a byte of data at the $R \times D$ input.

The most popular serial protocol is the (in)famous RS232 and this is supported by the 8051 family. P3.0 is connected to a serial-to-parallel converter, while P3.1 is connected to a parallel-to-serial converter as in Figure 12.4.3. The rate at which data is shifted in or out is very important in RS232 transmission. One of the timer/counters is usually used to set the baud rate (i.e. bits/second of transmission).

The serial port uses two registers and one of the timers. SCON is used for setting up and monitoring the serial port, while SBUF

Figure 12.4.3

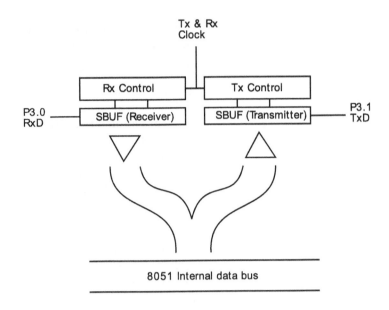

holds serial data. SBUF is physically two separate registers: one for holding data prior to serial transmission and one for accepting serial transmission and storing it before it is read by the CPU. Which register is accessed depends on whether the CPU performs a READ or WRITE to SBUF.

The control register SCON performs the following functions:

SM0	SCON.7	Serial Port control bit	Serial port mode bit 0
SM1	SCON.6	Serial Port control bit	Serial port mode bit 1
SM2	SCON.5	Serial Port control bit	Set by software to disable reception of frames for which bit 8 is zero
REN	SCON.4	Receiver Enable	1 = reception enabled 0 = reception disabled
TB8	SCON.3	Transmit bit 8	(In mode 2 or 3) is the 9th bit to be transmitted.
RB8	SCON.2	Receive bit 8	(In mode 2 or 3) is the 9th bit to be received.
TI	SCON.1	Transmit Interrupt flag	1 = transmit buffer empty (cleared by software)
RI	SCON.0	Receive Interrupt flag	1 = byte received into receive buffer (cleared by software)

The flexibility of RS232 is the cause of the complexity of this register. It is capable of operating in four modes according to the value of SM0 and SM1:

SM0	SM1	Mode	Function	Baud rate
0	0	0	Shift register I/O expansion	fOSC/12
0	1	1	8 bit UART	variable
1	0	2	9 bit UART	fOSC/32 or fOSC/64
1	1	3	9 bit UART	variable

- **Mode 0**
  The T × D pin outputs the shift register clock (at fOSC/12) while the R × D receives and transmits the 8 bits of data. The CPU is 'told' to expect incoming data when RI is set to 0 and REN is set to 1.
- **Mode 1**
  Data is handled in a conventional 1 start, 8 data, 1 stop RS232 style. The 10 bits are transmitted via T × D and received via R × D.
- **Mode 2**
  This mode allows the use of parity since the 9th data bit is handled by TB8 or RB8. So, in all, 11 bits are handled in the order 1 start, 8 data, RB8 (or TB8), 1 stop. It takes more effort to use since the programmer has to work out the parity of the data byte and then adjust the TB8 (or analyse the RB8) bit accordingly. This mode tends to be used for multiprocessor communication since it operates at $f_{OSO}/32$ or $f_{OSC}/64$ as set by the baud rate doubling bit PCON.7 (SMOD). (More of this in the following section on baud rate generation.)
- **Mode 3**
  Mode 3 is the same as mode 2 except that the baud rate is variable.

Modes 1 and 3 are used for RS 232 communication:

$$\text{Mode 1 for } 1 + 8 + 1$$
$$\text{Mode 3 for } 1 + 8 + P + 1$$

Baud rate generation

RS 232 conventionally uses one of a limited range of bit rates. In most cases, it is generated from the system clock and uses one of the timers. A common method is to use Timer 1 in its auto reload mode. To recap, this is a mode whereby TH1 is loaded into TL1 every time TL1's register overflows to 00. Before we detail some commonly used frequencies and their timer load values, there is yet one more register to explain. This is PCON. PCON is the **power control** register, and the 8051 only uses three of its bits for internal functions and two bits as general purpose flags which the programmer can utilize:

SMOD	PCON.7	Double baud rate. When SMOD = 1 a double baud rate is generated if Timer 1 is used and serial mode 1, 2 or 3 is selected
GF1	PCON.3	General purpose user flag
GF0	PCON.2	General purpose user flag
PD	PCON.1	Power Down. PD = 1 activates Power Down mode until the processor is reset. The clock is stopped and minimal power is consumed. Only data held in the on-chip RAM is preserved, but all other settings are lost
IDL	PCON.0	Idle. IDL = 1 disconnects the clock from the CPU but not from the interrupt,

timer or serial ports. All registers and settings are preserved. The processor can be woken up either by an interrupt or a full reset

Baud rate generation using Timer 1:

Baud rate	Frequency (MHz)	PCON.7 SMOD	Timer TH1
300	6	0	CC
300	11.059	0	A0
300	12	1	30
300	16	0	75
600	6	0	E6
600	11.05	0	D0
600	12	1	98
600	16	0	BB
1200	6	0	F3
1200	11.059	1	D0
1200	12	0	E6
1200	16	0	DD
2400	11.059	0	F4
2400	12	0	F3
2400	16	1	DD
4800	11.059	0	FA
4800	12	1	F3
9600	11.059	0	FD
9600	12	1	F9
19200	11.059	1	FD
19200	12	1	FD

Mode 0 example:

```
; Mode 0: serial data exits and enters through the R x D pin.
; The T x D pin outputs the shift clock. In mode 0, 8 bits are
; transmitted/received starting with the least significant bit
; The baud rate is fixed to 1/12 the oscillator frequency
;
    org   00h
    mov   scon, #00h    ; set up for mode 0
loop:
    mov   sbuf, #0aah   ; transmit aah
    jnb   ti,$          ; wait until done
    clr ti              ; clear transmit flag
    jmp loop            ; repeat again
    end
```

Mode 1 example:

```
; This program transmits hex value 'aa' continuously across
; the serial port of an 8051 in mode 1. It uses Timer 1 at 1200 baud
;
    org   00h
    mov   scon, #01000000B  ; set serial port for
                            ; mode 1 operation
    mov   tmod, #20h        ; set timer 1 to auto reload
    mov   th1, #0ddh        ; set reload value for
                            ; 1200 baud at 16 MHz
    setb  tr1               ; start timer 1
    clr   ti
```

```
loop:
    mov     sbuf, #0aah         ; transmit 'aa' hex out on T×D line
    jnb     ti,$                ; wait until done
    clr     ti                  ; clear transmit flag ready to go
                                ; again
    jmp     loop                ; repeat again
    end
```

Mode 2 example:

```
; A program to transmit the hex value 'aa' continuously out of
; the serial port of in mode 2 at 1/32 the oscillator frequency
;
    org     00h

    mov     scon, #10000000Bh   ; set up for mode 2
    setb    smod                ; baud rate equals 1/32 osc. freq
    clr     ti                  ; ready to transmit
loop:
    mov     sbuf, #0aah         ; transmit 'aa' hex
    jnb     ti,$                ; wait until done
    clr     ti                  ; clear transmit flag
    jmp     loop                ; repeat again

    end
```

Mode 1 example:

```
; This program continuously receives a byte entering
; the serial port pin R×D and puts the data out on port 1.
;
    org     00h
    jmp     main

    org     23h                 ; starting address of serial interrupt
    jmp     serial_isr

main:
    mov     scon, #50h          ; set up serial port for mode 1
                                ; with receive enabled
    mov     tmod, #00100000B    ; set up timer 1 as
                                ; auto-reload 8-bit timer
    mov     th1, #0ddh          ; baud rate equals 2400 baud at
                                ; 16 MHz
    setb    smod                ; set the double baud rate bit
    mov     ie, #10000001B
    setb    tr1                 ; start timer 1
    clr     ri                  ; ensure receive interrupt flag
                                ; is clear
loop:
    jmp     loop                ; endless loop
                                ; (unless interrupt occurs)
serial_isr:                     ; serial interrupt service routine
    mov     p1, sbuf            ; move the data to port
    clr     ri                  ; clear the ri bit
    reti                        ; return to the main program

    end
```

12.5 Analogue-to-digital and digital-to-analogue conversion

Microprocessors operate with digital quantities, but most real-world parameters are analogue and are continuously variable elements. To cope with this difference, there are a large number of electronic devices which convert between the two measurement systems: digital-to-analogue conversion (DAC) and analogue-to-digital conversion (ADC).

This section looks into the theory and application of DACs and ADCs – how they work and how they can be used.

Digital-to-analogue converters (DACs)

A digital-to-analogue converter takes a number of digital inputs and converts them into an analogue voltage as shown in Figure 12.5.1.

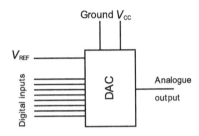

Figure 12.5.1 *The digital-to-analogue converter*

The reference voltage V_{REF} determines the range over which the analogue voltage is output. Usually, for an 8-bit DAC, 00000000 to 11111111 inputs would map to 0 V to V_{REF} V output. Some DACs use the supply voltage V_{CC} as the reference, although this is not the best of choices since any noise on the supply then affects the value of the output voltage.

The basic structure of a digital-to-analogue converter is a device called the R–$2R$ ladder. Figure 12.5.2 shows an R–$2R$ ladder as part of an op amp circuit.

Each switch from left to right can cause an output voltage V_{OUT}. The voltage output doubles for each switch. It is the ratio of the resistors which is important, not their actual value. When they are fabricated onto a piece of silicon, they can be matched to a high degree of accuracy.

The sequence of diagrams in Figure 12.5.3 show how the output is related to a different proportion of the reference voltage.

There are several categories of error which are attributable to DACs. These are:

- **Monotonicity** (Figure 12.5.4)

  When the input to the DAC is increased by 1 LSB steps, the output of the DAC should also increase by equal steps. This is the correct situation: the DAC produces ONE output for the input. Hence the MONO in monotonic. However, if, due to errors in the bit weighting, the output of the DAC decreases for any step, then the DAC is said to be non-monotonic.

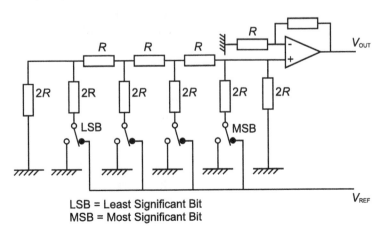

LSB = Least Significant Bit
MSB = Most Significant Bit

Figure 12.5.2 *The R–2R ladder*

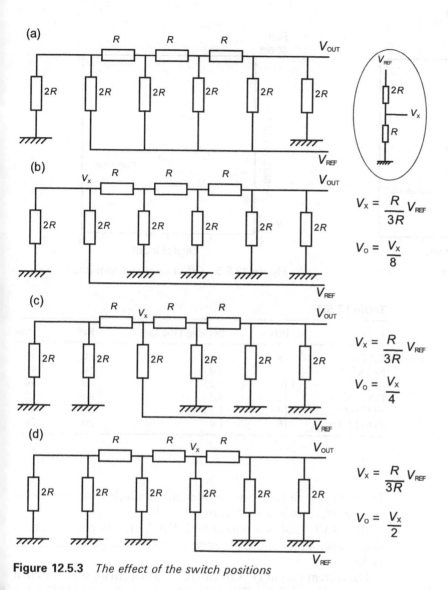

Figure 12.5.3 *The effect of the switch positions*

- **Offset error and gain error** (Figure 12.5.5)
  The output should normally be zero for zero input (although there are applications where this may deliberately be made not so). In some packages, offset voltages in the switches may create a small output. A gain error would exist if the output does not reach its maximum full scale output for the full scale range of the input.
- **Non-linearity** (Figure 12.5.6)
  This is the maximum amount which is expressed either as a percentage of the full-scale or as a fraction of an LSB by which any point of the transfer characteristic deviates from the ideal straight line passing through zero and the full scale output.

There is finite time before the output is stable due to the propagation delay of the interelectrode capacitances. This is not an error condition, but a factor of the DAC which must be taken into consideration when selecting a device. The other main factor is usually the resolution, or number of digital bits in the input. Table 12.5.1 shows a selection of DACs.

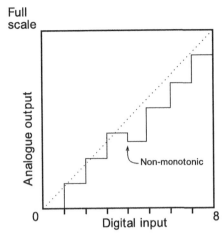

Figure 12.5.4 *Monotonicity*

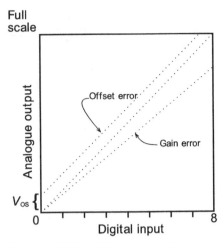

Figure 12.5.5 *Offset error and gain error*

Table 12.5.1

Type	Bits	Settling time (μs)	Ref	Pins
DAC08	8	0.15	ext	16
MAX505	8	6.0	ext	16
AD561	10	0.25	int	16
DAC667	12	4.0	int	28
AD7534	14	1.5	ext	20
TDA1541A	16	1.0	int	28

The voltage reference can be internal, in which case a precision voltage reference is provided, or external.

The MAX 7224 is a typical 8-bit DAC, and its package and a simple microprocessor application is shown in Figures 12.5.7 and 12.5.8.

The current output DAC08 can be used in a circuit similar to that shown in Figure 12.5.9. The main difference is that the DAC08 has a current output which has to be converted into a voltage. This can be done by a simple resistor or by an Operational Amplifier which gives the opportunity to produce gain.

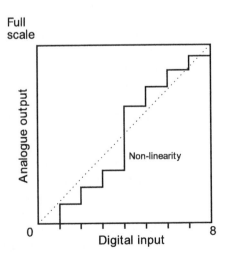

Figure 12.5.6 *Non-linearity*

Microprocessors 291

Figure 12.5.7 *The MAX 7224*

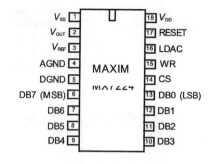

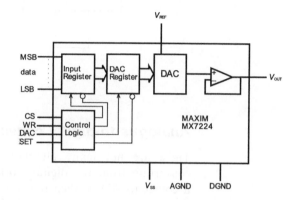

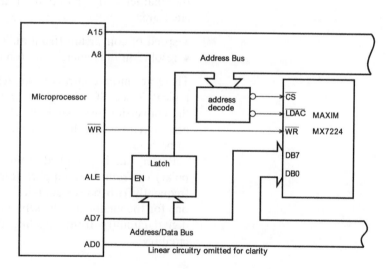

Figure 12.5.8 *The MAX7224 in a typical application*

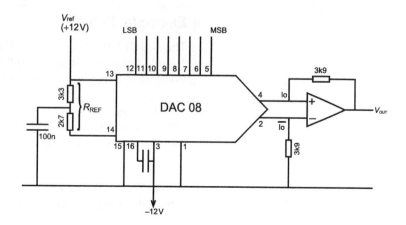

Figure 12.5.9 *The DAC08*

Figure 12.5.10 *An ADC*

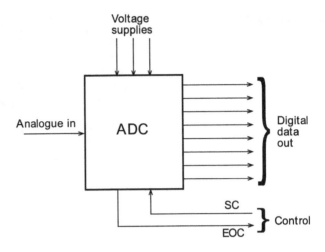

Analogue-to-digital converters

There are not many alternatives to the methods in use for converting from the digital world into the analogue, but when it comes to ADCs, then the conversion method used depends on the characteristics required of the converter. The main parameters are:

- speed of conversion (from millisecond to sub-microsecond);
- resolution (anywhere from 6-bit to 16 or more bits).

There are more controls to an ADC, and Figure 12.5.10 shows a typical device. SC (Start Convert) is an external command to start the conversion process; EOC (End Of Conversion) is a signal to other circuitry that the conversion is complete and the data is ready to be read.

Power can be supplied from one single positive voltage (unipolar) or from dual supplies (bipolar). As with DACs, there are frequently two power supplies – one for the digital circuitry and one for the analogue. The separation is used to prevent the noisier digital circuitry from affecting the (hopefully cleaner) analogue circuitry.

> ### Example 12.5.1
>
> The ADC1121 is a 12-bit ADC with a basic resolution of 1.2 mV. What would be the full scale capability of the device? What attenuator would be required to measure 0–25 V?
>
> The LSB is 1.2 mV and a 12-bit counter can output from 000000000000 to 111111111111 which is 0 to 8191. The maximum output would be:
>
> $$V_{OUT} = 1.2\,\text{mV} \times 8191 = 9.8292\,\text{V}$$
>
> If the input is actually required to measure 0 to 25 V, then the attenuator ought to be 25 : 9.8292.

Problems 12.5.1

(1) The temperature in a kettle is measured and an 8-bit ADC used to present a digital version of the temperature as it rises from room temperature to 100 °C. If the converter is calibrated to convert from 0 °C to 100 °C over its full digital range, what is the basic resolution of the system?

(2) What would be the minimum number of bits necessary to measure 0–30 V with an accuracy of 0.2%?

(3) An 8-bit ADC is just insensitive to $0.7\,mV_{RMS}$ of sinusoidal hum. What would be the Full Scale of this device?

(4) A pick and place robot has an arm of effective radius 0.5 m. The angular position is measured by an analogue potentiometer as the arm swings around. How many bits of resolution are required if the arm is to be capable of positioning an object with 1 mm accuracy?

Conversion methods

There are many methods of conversion, and this section will explain how they work and the advantages of using them. A few real-life applications will be presented at the end of the section, but with these highly integrated devices, the application is usually dictated by the manufacturers' notes.

The flash converter

This is the fastest method of conversion and is sometimes called parallel encoding. It is as fast as the switching delay in a comparator. The main problem can be the size and propagation delay introduced by the logic array which converts a one of n input into a binary output.

In the example of Figure 12.5.11, the logic array has 8 inputs (with 8 comparators) just to provide a 3-bit output. There are 8-bit flash converters (= 256 comparators) which can convert an input in a few tens of nanoseconds.

Count-up/count down converter

This is a feedback method where the output of a DAC compares the converted output with the input and ramps an up/down counter to eliminate any difference between the two (Figure 12.5.12).

The polarity of the up/down control determines whether the counter counts up or counts down. The result of the comparison between the DAC output and the actual analogue input is used to drive the up/down counter. Only the initial count is long. The device has good tracking, but does oscillate by ±1 bit about the final value.

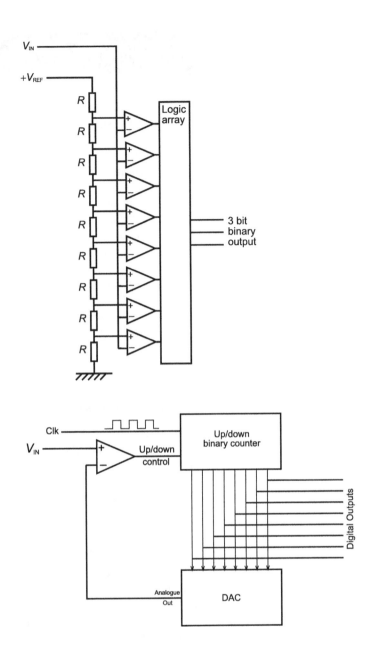

Figure 12.5.11 *The flash converter*

Figure 12.5.12 *The tracking converter*

Successive approximation

This has a moderate to high resolution (up to 16 bits) and moderate to high speed 1 ms to 250 ms. It is used on the popular ranges of ADCs such as the ADC0803/4/8/9/17, ZN449, MAX186 and AD574.

They require an internal or external clock, and for each clock pulse a binary search proceeds in a similar way that an old fashioned chemical balance can find the weight of a substance by adding or removing weights from a pan. Initially, all of the output bits are set to 0, then each bit in turn is set to 1. If this causes the output to exceed the input, then that bit is set back to 0, otherwise it is kept to 1. An 8-bit successive approximation converter will require 8 pulses to convert the input to a digital value. The faster the clock, the faster the conversion.

Integrating ADCs

The integration technique is comparatively slow, but can produce very accurate results. The conversion time is of the order of milliseconds rather than microseconds. It is integrating converters that appear in digital voltmeters where accuracy rather than speed of conversion is more important.

Single slope integrating ADC

In this ADC, a ramp generator is constructed from a current source and a capacitor. The voltage on the capacitor is allowed to ramp up until it is the same value as the input voltage. At the same time as the ramp is started, a clock is started and the clock pulses counted into a register. The longer it takes for the ramp to reach the input voltage, the more pulses will be stored. At the end of the conversion (ramp voltage being the same as the input voltage) the capacitor is discharged and the register contents are transferred to an output buffer before the whole procedure starts again. In practice, it is difficult to maintain the stability and accuracy of the current source and capacitor, so the technique is not often used.

Dual slope integrating ADC

This is the form of the integrating ADC which is accurate, repeatable and reliable and is used in many precision digital multimeters. It extends the principle of the single slope and at the same time eliminates most of the single slope converter's problems.

Firstly, current proportional to the input level charges a capacitor for a fixed time. Then the capacitor discharges at a constant current until the voltage reaches zero again. The time taken to discharge the capacitor is proportional to the input level and is used to gate a counter driven at a fixed frequency. The final count is proportional to the input level.

A variant of the dual slope converter is the quad slope converter. This operates in a similar way except that there is an 'auto-zero' cycle where the input is connected to the analogue 0 V line and a dual slope measurement taken. This helps cancel offset errors by

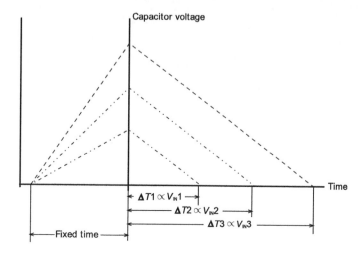

Figure 12.5.13 *Dual slope inegrating ADC*

Figure 12.5.14 *The MAX186 ADC*

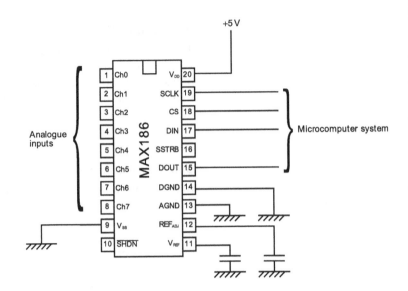

subtracting any voltages measured in the auto-zero cycle from the actual measurement cycle.

Serial input/output converters

This does not refer to the conversion technology but rather to the interface methods used. If speed is not the absolute criterion, then designers will often use a DAC or ADC which can interface to a computer or microcontroller via a serial bus. This helps keep down the circuit board area and can reduce the size of the final product. For example, Maxim produce the MAX186 which is an 8-channel (i.e. 8 input) ADC with 12-bit resolution and internal voltage reference and clock which communicates with a computer over a 4-wire serial interface (Figure 12.5.14). This provides all the control, status and data information that the computer requires. All this in a 20-pin IC. This would make it ideal for a data logger type of application.

Appendix A Number systems

Computers deal in binary quantities – i.e. a 'low' voltage (0–0.8 V) is a logic '0' while a 'high' voltage (3.4–5.0 V) is a logic '1'. Humans work in a decimal system, and as such you probably learnt your maths as Hundreds, Tens and Units (HTU). Mathematically, we could express this as 10^2, 10^1, 10^0, so a sum would be:

$$
\begin{array}{ccc}
H & T & U \\
10^2 & 10^1 & 10^0 \\
\hline
4 & 6 & 2 \\
+\quad 1 & 8 & 3 \\
\hline
6 & 4 & 5 \\
\end{array}
$$

Note that adding 6 tens to 8 tens generates a 'carry', i.e. it overflows the capability of the Tens column. The same happens in the **binary** system. We do not have Hundreds, Tens and Units, but we do have powers of 2:

$$
\begin{array}{cccc}
8 & 4 & 2 & 1 \\
2^3 & 2^2 & 2^1 & 2^0 \\
\end{array}
$$

A decimal system has ten symbols (0, 1, 2, 3, 4, 5, 6, 7, 8, 9) while the binary system has two (0, 1).

Binary to decimal conversion

A binary number can simply be converted to decimal by noting the position and value of the 1s:

$$
\begin{array}{cccccccc}
2^7 & 2^6 & 2^5 & 2^4 & 2^3 & 2^2 & 2^1 & 2^0 \\
128 & 64 & 32 & 16 & 8 & 4 & 2 & 1 \\
\end{array}
$$

$0\ 1\ 1\ 0\ 0\ 1\ 0\ 1 = 64 + 32 + 4 + 1 = 101_{10}$

$0\ 1\ 1\ 1\ 1\ 0\ 0\ 0 = 64 + 32 + 16 + 8 = 120_{10}$

$0\ 0\ 0\ 1\ 0\ 1\ 0\ 0 = 16 + 4 = 20_{10}$

Problem A1.1

Convert the following binary numbers to decimal:

(a) 10110
(b) 11000011
(c) 10100101
(d) 1000101
(e) 11010111
(f) 01111111

Decimal to binary conversion

A common way of converting decimal numbers to binary is to use the 'repeated division by 2' method. For example, convert 207_{10} into binary:

```
2) 207
2) 103  1
2)  51  1
2)  25  1
2)  12  1
2)   6  0
2)   3  0
     1  1
```

The initial number is repeatedly divided by 2 until either a '1' or a '0' is left. At each stage, the remainder is written down, i.e. 207 divided by 2 is 103 r 1 etc. The answer is then read from bottom to top:

$$207_{10} = 11001111_2$$

Problem A1.2

Convert the following decimal numbers to binary:

(a) 112
(b) 69
(c) 29
(d) 193
(e) 87
(f) 237

Hexadecimal notation

Computers work in binary quantities, and also handle words of various sizes. These could be 4, 8, 16, 32 or even 64 bits wide at a time. In this book, we are mostly concerned with 8-bit processors. 8 bits are known as a byte, while 4 bits are known as a nibble. 4 bits contain 16 possible binary numbers, so it has become convenient to use a numbering system called hexadecimal. In this system there are 16 single symbols: 0, 1, 2, 3, 4, 5, 6, 7, 8, 9, A, B, C, D, E, F. Our 'HTU' sum has now become:

$$16^3 \quad 16^2 \quad 16^1 \quad 16^0$$
$$4096 \quad 256 \quad 16 \quad 1$$

Decimal to hexadecimal conversion

Given a number such as 237, it will be found that 237 divided by 16 is 14 r 13:

14 has a hexadecimal value of E

13 has a hexadecimal value of D

So, $237_{10} = ED_{16}$.

Problem A1.3

Convert these decimal numbers into hexadecimal:

(a) 63
(b) 99
(c) 180
(d) 200

Hexadecimal to decimal conversion

Conversion of a number such as 7A requires the use of both number systems:

A has a decimal value of 10

7 in the 16^1 column has a decimal value of $7 \times 16 = 112$

So $7A_{16} = 112_{10} + 10_{10} = 122_{10}$.

Problem A1.4

Convert these hexadecimal numbers into decimal:

(a) 22
(b) C9
(c) 1B
(d) F3

Of course, the simple way to convert is to use the Convert facility available on most calculators. However, there is no reason for not knowing the mechanism behind the conversion principle.

Logic

There are four main logical fuctions: AND, OR, NOT and EXCLUSIVE OR. They are defined as follows:

AND — The output is true if and only if all inputs are true $X = A \cdot B$. is the AND function

OR — The output is true if any of the inputs are true $X = A + B$ + is the OR function

NOT — The output is true if the output is false $\overline{X} = \overline{A}$ is the NOT function

EXCLUSIVE OR — The output is true if only one of the inputs is true $X = A\ B$ is the EXCLUSIVE OR function

The following examples show the effects of two 8-bit numbers being combined:

```
AND  00111010
     11110110
     --------
     00110010
```

Only where both digits are a 1 will the output be a 1.

Problem A1.5

Complete the following sums:

```
11000011   10011001   10111110   00100100
11010111   01110100   11000010   00111101
--------   --------   --------   --------
```

```
OR   11001100
     10101010
     --------
     11101110
```

The output is a 1 if any of the inputs is a 1.

Problem A1.6

Complete the following sums:

```
10000001   11001100
00001100   10100101
--------   --------
```

Applications of logic functions

It is sometimes necessary to SET (to 1) or RESET (to 0) any single bit of an 8-bit byte. This is done by a process called **masking**. For example, if it is necessary to set bit 3 of a byte, then that byte should be OR'd with:

00001000 (note that bit 0 is the Least Significant Bit – LSB)

This is known as the **mask byte**. When any two numbers are OR'd together, a 1 in any position causes a 1 in the output. This mask byte would cause bit 3 to be set to 1. That is, if the byte was 11000011:

$$\begin{array}{rl} \text{mask} & \underline{00001000} \quad \text{OR} \\ \text{result} & 11001011 \end{array}$$

Problem A1.7

What mask byte would be used to set:

(a) bits 0 and 7;
(b) bits 2 and 4;
(c) bit 6;
(d) the lower nibble?

If a bit needs to be RESET, the AND function is used. ANDing a number with 11110111 would cause bit 3 to be RESET, and all others to be left unchanged. This is because:

0 ANDed with 1 = 0 and 1 ANDed with 1 = 1

0 ANDed with 0 = 0 and 1 ANDed with 0 = 0

For example, to reset bit 7 – AND with 01111111:

$$\begin{array}{l} 11000110 \\ \underline{01111111} \quad \text{AND} \\ 01000110 \end{array}$$

If the supplied byte was 11000110 then the result would be 01000110, i.e. unchanged apart from bit 7.

Problem A1.8

What mask would be used to clear:

(a) the upper nibble;
(b) the lower nibble?

American Standard Code for Information Interchange (ASCII)

Whenever characters, numbers and other codes need to be sent from one device to another, it is frequently encoded into ASCII. Originally, this was a 7-bit code, but now has been extended into an 8-bit code to cover a greater variety of symbols. Most of the codes are directly for symbols, but the first 32 are used as control characters. They control things like serial transmission of data, or printers, or file exchange.

CH	HEX	DEC	CH	HEX	DEC	CH	HEX	DEC
NUL	00	00	+	2B	43	V	56	86
SOH	01	01	,	2C	44	W	57	87
STX	02	02	-	2D	45	X	58	88
ETX	03	03	.	2E	46	Y	59	89
EOT	04	04	/	2F	47	Z	5A	90
ENQ	05	05	0	30	48	[	5B	91
ACK	06	06	1	31	49	\	5C	92
BEL	07	07	2	32	50	]	5D	93
BS	08	08	3	33	51	^	5E	94
HT	09	09	4	34	52	_	5F	95
LF	0A	10	5	35	53	`	60	96
VT	0B	11	6	36	54	a	61	97
FF	0C	12	7	37	55	b	62	98
CR	0D	13	8	38	56	c	63	99
SO	0E	14	9	39	57	d	64	100
SI	0F	15	:	3A	58	e	65	101
DLE	10	16	;	3B	59	f	66	102
DC1	11	17	<	3C	60	g	67	103
DC2	12	18	=	3D	61	h	68	104
DC3	13	19	>	3E	62	i	69	105
DC4	14	20	?	3F	63	j	6A	106
NAK	15	21	@	40	64	k	6B	107
SYN	16	22	A	41	65	l	6C	108
ETB	17	23	B	42	66	m	6D	109
CAN	18	24	C	43	67	n	6E	110
EM	19	25	D	44	68	o	6F	111
SUB	1A	26	E	45	69	p	70	112
ESC	1B	27	F	46	70	q	71	113
FS	1C	28	G	47	71	r	72	114
GS	1D	29	H	48	72	s	73	115
RS	1E	30	I	49	73	t	74	116
US	1F	31	J	4A	74	u	75	117
spc	20	32	K	4B	75	v	76	118
!	21	33	L	4C	76	w	77	119
"	22	34	M	4D	77	x	78	120
#	23	35	N	4E	78	y	79	121
$	24	36	O	4F	79	z	7A	122
%	25	37	P	50	80	{	7B	123
&	26	38	Q	51	81	|	7C	124
'	27	39	R	52	82	}	7D	125
(	28	40	S	53	83	~	7E	126
)	29	41	T	54	84	DEL	7F	127
*	2A	42	U	55	85			

Appendix B: MCS51 Instruction set

Notes

Rn	Register R7-R0 of currently selected register bank
direct	8-bit internal data location's address
@Ri	Internal RAM location 0-255 addressed indirectly through R7-R0
#data	8-bit constant included in instruction
#data16	16-bit constant included in instruction
addr16	16-bit destination address (used in LCALL and LJMP)
addr11	11-bit destination address (used in ACALL and AJMP)
rel	Signed (2's complement) 8-bit offset byte (SJMP)
bit	Direct addressed bit in internal data RAM or Special Function Register

Arithmetic operations

ADD	A,Rn	Add register n to A
ADD	A,direct	Add direct byte to A
ADD	A,@Ri	Add indirect RAM to A
ADD	A,#data	Add immediate data to A
ADDC	A,Rn	Add with carry Register n to A
ADDC	A,direct	Add with carry direct byte to A
ADDC	A,@Ri	Add with carry indirect RAM to A
ADDC	A,#data	Add with carry immediate data to A
SUBB	A,Rn	Subtract with borrow Register n from A
SUBB	A,direct	Subtract with borrow direct byte from A
SUBB	A,@Ri	Subtract with borrow indirect RAM from A
SUBB	A,#data	Subtract with borrow immediate data from A
INC	A	Increment Accumulator
INC	Rn	Increment Register n

INC	direct	Increment direct byte
INC	@Ri	Increment indirect RAM
DEC	A	Decrement Accumulator
DEC	Rn	Decrement Register n
DEC	direct	Decrement direct byte
DEC	@Ri	Decrement indirect RAM
INC	DPTR	Increment Data Pointer
MUL	AB	Multiply A and B
DIV	AB	Divide A by B
DA	A	Decimal Adjust Accumulator

Logical operations

ANL	A.Rn	AND Register n to A
ANL	A,direct	AND direct byte to A
ANL	A,@Ri	AND indirect RAM to A
ANL	A,#data	AND immediate data to A
ANL	direct,A	AND A to direct byte
ANL	direct,#data	AND immediate data to direct byte
ORL	A.Rn	OR Register n to A
ORL	A,direct	OR direct byte to A
ORL	A,@Ri	OR indirect RAM to A
ORL	A,#data	OR immediate data to A
ORL	direct,A	OR A to direct byte
ORL	direct,#data	OR immediate data to direct byte
XRL	A.Rn	Ex-OR Register n to A
XRL	A,direct	Ex-OR direct byte to A
XRL	A,@Ri	Ex-OR indirect RAM to A
XRL	A,#data	Ex-OR immediate data to A
XRL	direct,A	Ex-OR A to direct byte
XRL	direct,#data	Ex-OR immediate data to direct byte
CLR	A	Clear A
CPL	A	Complement A
RL	A	Rotate A Left
RLC	A	Rotate A Left through Carry
RR	A	Rotate A Right
RLC	A	Rotate A Right through Carry
SWAP	A	Swap nibbles within A

Data transfer

MOV	A,Rn	Move Register n to A
MOV	A,direct	Move direct byte to A
MOV	A,@Ri	Move indirect RAM to A
MOV	A,#data	Move immediate data to A
MOV	Rn,A	Move A to Register n
MOV	Rn,direct	Move direct byte to Register n
MOV	Rn,#data	Move immediate data to Register n
MOV	direct,A	Move A to direct byte
MOV	direct,Rn	Move Register n to direct byte
MOV	direct,direct	Move direct byte to direct byte
MOV	direct,@Ri	Move indirect RAM to direct byte
MOV	direct,#data	Move immediate data to direct byte
MOV	@Ri,A	Move A to indirect RAM

MOV	@Ri,direct	Move direct byte to indirect RAM
MOV	@Ri,#data	Move immediate data to indirect RAM
MOV	DPTR,#data16	Load data pointer with a 16-bit constant
MOVC	A,@A+DPTR	Move code byte relative to DPTR to A
MOVC	A,@A+PC	Move code byte relative to PC to A
MOVX	A,@Ri	Move external RAM (8-bit) to A
MOVX	A,@DPTR	Move external RAM (16-bit) to A
MOVX	@Ri,A	Move A to external RAM (8-bit)
MOVX	@DPTR,A	Move A to external RAM (16-bit)
PUSH	direct	Push direct byte onto the stack
POP	direct	Pop direct byte from stack
XCH	A,Rn	Exchange register n with A
XCH	A,direct	Exchange direct byte with A
XCH	A,@Ri	Exchange indirect RAM with A
XCHD	A,@Ri	Exchange low-order digit indirect RAM with A

Boolean variable manipulation

CLR	C	Clear carry
CLR	bit	Clear direct bit
SETB	C	Set carry
SETB	bit	Set direct bit
CPL	C	Complement carry
CPL	bit	Complement direct bit
ANL	C,bit	AND direct bit to carry
ANL	C,/bit	AND complement of direct bit
ORL	C,bit	OR direct bit to carry
ORL	C,/bit	OR complement of direct bit
MOV	C,bit	Move direct bit to carry
MOV	bit,C	Move carry to direct bit
JC	rel	Jump if carry set
JNC	rel	Jump if carry not set
JB	bit,rel	Jump if direct bit set
JNB	bit,rel	Jump if direct bit not set
JBC	bit,rel	Jump if direct bit is set and clear bit

Program branching

ACALL	addr11	Absolute subroutine call
LCALL	addr16	Long subroutine call
RET		Return from subroutine
RETI		Return from interrupt
AJMP	addr11	Absolute jump
LJMP	addr16	Long jump
SJMP	rel	Short jump (relative addr)
JMP	@A+DPTR	Jump indirect relative to the DPTR
JZ	rel	Jump if A is zero
JNZ	rel	Jump if A is not zero
CJNE	A,direct,rel	Compare direct byte to A and jump if not equal

CJNE	A,#data,rel	Compare immediate to A and Jump if not equal
CJNE	Rn,#data,rel	Compare immediate to reg n and jump if not equal
CJNE	@Ri,#data,rel	Compare immediate to indirect and jump if not equal
DJNZ	Rn,rel	Decrement register n and jump if not zero
DJNZ	direct,rel	Decrement direct byte and jump if not zero
NOP		No operation

Index

Accumulator, 273
ASCII table, 302
Amplifier:
    bridges, 81
    cascode, 195
    class A, 63, 64
    class B, 63, 75, 78
    class C, 64, 80
    common base, 194
    commutating auto-zero,, 141
    crossover distortion 75
    Darlington pair, 70
    frequency response, 60
    high pass, 62
    isolation, 125
    long tailed pair, 52, 95
    low pass, 62
    r.f. operational amplifier 196
    transconductance, 87, 161
    transresistance, 87
    tuned amplifier, 59
Analogue to digital converters, 260, 292
    count up/down, 293
    dual slope, 295
    flash, 293
    integrating, 295
    serial, 296
    single slope, 295
    successive approximation, 294
Arithmetic and logic unit, 260
Assembler directives, 275
Asynchronous counters, 238
Avalanche diode, 10

Bandwidth (operational amplifiers), 111
Baud rate generation, 285
Binary to decimal conversion, 297
Boltzmann's constant, 126
Boolean logic, 212

Boost regulator, 38
Bridge rectifier, 7
Buck converter, 36
Buck-boost regulator, 38
Buffer amplifier, 105
Burst (popcorn) noise, 130
Butler overtone oscillator, 200
Byte, 259

Capacitors:
    coupling, 124
    decoupling, 121
    electrolytic, 27
    ripple suppression, 8
Cascode amplifier, 195
Ceramic resonators, 200
Charge pump, 40
Chopper stabilised amplifier, 140
Class A power amplifier, 63, 64
Class B power amplifier, 63, 75
Class C power amplifier, 64, 80
Clock circuits, 264
CMOS, 207, 218
Co-axial cable, 125
Cockroft-Walton voltage multiplier, 30
Colpitts oscillator, 165
Combinational logic, 221
Common mode rejection ratio, 53, 118
Commutating auto-zero amplifier, 141
Comparator, 18, 21, 142
    base-emitter, 55
    phase, 201
    window, 146
Constant current supplies, 32
    BJT, 33
    FET, 32
    Howland current source, 34
    IC voltage regulator, 34

Counters:
   asynchronous, 238
   synchronous, 249
Crossover distortion, 75
Crosstalk, 120, 124
Crystal oscillators, 197
Cuk converter, 35
Current limiting, 16

D-type latch, 240
Darlington pair, 16, 70
D.C. restoration, 30
Decimal to binary conversion, 298
Decimal to hexadecimal conversion, 299
Decoder, tone, 204
Decoupling capacitors, 121
Demodulator, FM, 202
Digital to analogue converters, 288
Differential amplifier, 109
Diode pump voltage level converter, 31
Direct memory space, 271
Distortion, 84
   harmonic, 85
   intermodulation, 86
Dual Slope ADC, 295
Dynamic RAM, 256
Dynamic Resistance, 164

EAROM, 258
EEPROM, 257
Edge triggered logic, 247
Electrostatic sensitivity, 219
EPLD, 229
EPROM, 229, 256

Fan-out, 218
Feedback, 41
   benefits of negative feedback, 46
   negative, 44
   positive, 43
   series current, 88
   series voltage, 93
   shunt voltage, 98
   shunt current, 96
Fetch-execute cycle, 259
Filter:
   active, 152
   adjustable-Q notch, 160
   sallen and Key, 155
   twin-T notch, 160
Flash ADC, 293
Flicker ($1/f$) noise 129

Flyback regulator, 39
FM demodulator, 202
Foldback, 28
Frequency synthesiser, programmable, 208

Gain bandwidth product, 113
Gray code, 223
Ground loops, 122

Hartley oscillator, 165
Harmonic distortion, 85
Heatsinks, 26
Hexadecimal to decimal conversion, 299
Hexadecimal notation, 298
Hickman voltage multiplier, 31
Howland constant current source, 34
Hybrid-PI equivalent circuit, 193

IEC qualifying symbols, 215
Inductive coupling, 124
Inductors, strip-line, 196
Input offset current, 117
Input offset voltage, 117
Instruction register, 260
Instrumentation amplifier, 133
Integrated circuit power amplifiers, 80
Integrating ADC, 295
Integrator, operational amplifier, 148
Interference, 120
Intermodulation distortion, 86
Intel 8051 Instruction Set, 303
Interrupts, 279
Interrupt service routine, 279
Inverting operational amplifier, 106

J-K flip flops, 246
Johnson noise, 126

Karnaugh map, 223

Latches, 238
Logic gates, 211
   NAND, 214
   NOR, 214
   NOT, 212
Logic minimisation techniques, 221
Logic operations, 300
Long tailed pair, 52, 95

Magnetic coupling, 124
Magneto resistor, 137
Master-slave logic, 247
MCS51 instruction set, 303
Mealy machine, 247
Memory:
   RAM, 255
   ROM, 256
Microprocessors, 259
   accumulator, 273
   arithmetic and logic unit, 260
   assembler directives, 275
   baud rate generation, 285
   byte, 259
   clock circuits, 264
   fetch-execute cycle, 259
   instruction register, 260
   Intel 8051 instruction set, 303
   interrupts, 279
   interrupt service routine, 279
   MCS51 instruction set, 303
   nibble, 259
   non volatile memory, 269
   power-on-reset, 263
   program counter, 260
   program status word, 273
   registers, 271
   special function register, 271
   timer/counters, 275
   UART 260
   vector addressing, 279
   watchdog timer, 263
Miller effect, 193
Mixer amplifier, 108
Moore machine, 247
Multivibrator:
   astable transistor, 182
   astable, '555', 190
   astable operational amplifier, 180
   monostable transistor, 185
   monostable, '555', 188

Negative feedback, 44
Nibble, 259
Non volatile memory, 269
NAND logic gate, 214
Noise:
   burst (popcorn), 130
   figure, 131
   flicker ($1/f$), 129
   Johnson noise, 126
   pink, 129
   shot, 129
   temperature, 132

   white, 126
Non-inverting operational amplifier, 105
NOR logic gate, 214
NOT logic gate, 212

Open collector output, 215
Open drain output, 266
Operational amplifiers, 102, 196
   bandwidth, 111
   buffer, 105
   chopper stabilised, 140
   common mode rejection ratio, 118
   commutating auto zero, 141
   differential, 109
   gain bandwidth product, 113
   input offset current, 117
   input offset voltage, 117
   instrumentation, 133
   integrator, 148
   inverting, 106
   mixer, 108
   non-inverting, 105
   slew rate limiting, 111
   virtual earth
Oscillators:
   astable '555', multivibrator, 190
   Butler overtone, 200
   Clapp oscillator, 199
   Colpitts oscillator, 165
   Hartley oscillator, 165
   phase shift ladder, 169
   quartz crystal, 197
   R-C logic gate, 245
   relaxation, 177
   triangle wave, 150
   tuned collector, 163
   twin-T, 175
   voltage controlled (PLL), 201, 203
   Wien bridge, 172
Overcurrent protection, 17

PAL, 228
PALASM (TM), 234
Peak inverse voltage (PIV), 10
Peak voltage clamp, 30
Phase comparator, 201
Phase lock loop (PLL), 201, 203
   capture range, 204
   lock range, 204
   voltage controlled oscillator, 203
Phase shift ladder oscillator, 169
Photo conductive cell, 137
Piezoelectric effect, 197

Pink noise, 129
PLA, 228
Platinum resistance thermometer, 137
Positive feedback, 43
Power amplifiers, 57
Power-on-reset, 263
Program counter, 260
Program status word, 273
Programmable frequency synthesiser, 208
Progammable logic, 225, 253
    PAL, 228
    PLA, 228
    PROM, 229
PROM, 229
Pulse width modulation (PWM), 35

Quartz, 197
Quine-McKluskey minimisation, 223

Ramp generator, 150
Rectification, 5
Rectifier bridge, 7
Registers, 271
Regulation, 4
Relaxation oscillator, 177
R.F. bands, 192
RMS quantities, 2
RS232, 283
Rubber zener (V_{BE} multiplier), 76
Sallen and Key active filters, 155
Schmitt trigger, 216
Screen, high-mu, 122
Sequential logic, 238
Serial communication, 283
Serial input/output ADC converters, 296
Set-reset latch, 238
Shift register, 252
Shot noise, 129
Single slope ADC, 295
Slew Rate Limiting 111
SSI small scale integration, 217
Special function register, 271
State transition diagram, 248
Static RAM, 255
Strain gauge, 137
Subtractor amplifier, 109
Successive approximation ADC, 294
Switch debouncer, 240
Switch logic, 211

Switched mode power supplies, 34
    buck converter, 36
    buck-boost regulator, 38
    boost regulator, 38
    charge pump, 40
    Cuk converter, 35
    flyback regulator, 39
Synchronous counter, 249

Thermal resistance, 25
Three terminal voltage regulators, 22
Timer/counters, 275
Timer '555', 187
Tone decoder, 202, 204
Totem pole output, 214
Transconductance amplifier, 161
Transformers, 1
    R.F., 195
Triangle-wave generator, 150
Tri-state logic, 216
Truth table, 213
Tuned amplifier, 59
Tuned collector oscillator, 163
Twin-T notch filter, 160
Twin-T oscillator, 175

UART, 260

Vector addressing, 279
Virtual earth, 105
Volt-amperes, 2
Voltage multipliers:
    Cockroft-Walton, 30
    diode pump, 31
    doubler, 30
    Hickman, 31
Voltage regulators (IC), 22

Watchdog timer, 263
Wheatstone Bridge, 136
White Noise, 126
Wien Bridge Oscillator, 172
Window Comparator, 146
Wired-Or Logic, 146

Zener Diode, 10

For Product Safety Concerns and Information please contact our EU representative GPSR@taylorandfrancis.com Taylor & Francis Verlag GmbH, Kaufingerstraße 24, 80331 München, Germany

Printed and bound by CPI Group (UK) Ltd, Croydon, CR0 4YY
08/06/2025
01897007-0018